Maths, Art, or Magic?

A user guide to the art of radio station scheduling and programming

by

Robin Prior

www.mathsartormagic.net

To Dr Andrew Economos: Thank you for allowing me to work with you and RCS for so many decades. It has brought me joy, challenges, and the satisfaction that I was doing something of value.

To Athol Wesselink: Your friendship and ongoing encouragement have been part of my journey. Your persistent nudges to document our work have been a constant reminder of the value of our experiences. Thank you for being a part of this.

To Colin and Nateekan, who devoted many weeks of attention to clarifying the text and addressing missing details. Thank you, the book is much better for your efforts.

And, above anyone else, and most importantly:
To my wife, Marilyn Prior, who has been my unwavering support. You accepted my passion for radio and your role as the rock of our relationship. You graciously accepted RCS as a second lover in our marriage, demanding far too much of my time. More, I could not wish for.

Contents

Foreword

It gives me great pleasure to commend Robin Prior on his remarkable book Maths, Art, or Magic? Radio Station Scheduling and Programming.

"When I founded Radio Computing Services and created Selector, our vision was to bring mathematical structure and efficiency to the art of radio programming. Over the decades, Selector and its successors have been used by broadcasters worldwide, but what has always distinguished the best programmers is not simply their mastery of the software, but their ability to blend science with creativity and intuition.

"Robin embodies this rare balance. During his more than forty years with RCS, including his leadership as RCS Africa VP and his pioneering work in developing and implementing digital playout systems, Robin consistently demonstrated deep technical knowledge, sound judgment, and a unique sensitivity to what makes radio truly connect with listeners. "His book distils that wealth of experience into a guide that is as practical as it is inspiring.

By exploring the interplay of mathematics, artistry, and what can only be described as magic, Robin has provided broadcasters and programmers with a thoughtful framework to understand not just how scheduling works, but why it matters. His 16-step methodology, his attention to the listener's emotional journey, and his integration of modern media strategies make this work both timely and timeless. "I am proud to have worked with Robin and to see his contributions to the field recognized in this way.

"Maths, Art, or Magic? will serve as an invaluable resource for radio professionals everywhere, and a fitting testament to the knowledge and passion Robin has shared with our industry."

With appreciation and respect,

Andrew Economos, PhD

Founder, Radio Computing Services

Introduction

If you choose to seek blame for how many of the major radio stations' music in Africa is scheduled by smart computer software, blame Robin Prior.

I have spent many decades training radio station programmers and schedulers worldwide, particularly in Africa, including state broadcasters such as the SABC and Zimbabwe Broadcasting Corporation, on Selector and GSelector.

These include virtually every commercial station and a significant number of community stations in South Africa and major groups in Kenya, Uganda, Tanzania, Nigeria, Namibia, and Botswana.

Over the years, I have consistently applied these essential philosophies every time I begin training. I realized that it was appropriate to document all this so that I could hopefully speed up our training process in the future.

Full disclosure: I am not, and never have been, responsible for programming a radio station myself; though having worked with some of the best radio programmers on Earth at streamlining their scheduling systems.

I'd like to pass on some of their knowledge to you about the Maths, Art, and Magic of Radio and TV Music Scheduling and Programming.

Further disclosure: I have some heroes in radio management and programming from whom I draw inspiration and joyfully quote.

Brad March is a consummate and highly respected radio man with a career-long succession of firsts in the Australian broadcasting industry. Reading about his exploits often opens my eyes wide. Brad's career has included everything from being an announcer to Group Content Consultant for Austereo. As Managing Director, he and his team launched Austereo on the ASX. He is presently the MD of Marchmedia, a talent management company. I greatly respect the impact that Brad has had on Australian and worldwide radio. I am grateful that he responded to my queries and concerns about the book, and I have quoted him verbatim extensively.

Then there is Randy Michaels. Controversial? Yes. But he is the man who took Clear Channel from a group of 200 to 1,200 radio stations. Randy is the most passionate and competitive radio man I have ever met.

If you ever compete directly with Randy, be prepared for unconventional and off-the-wall tactics; he practices radio guerrilla warfare to the extreme.

Respect him and fear him.

Randy is famously known for his frequent rant at the top of his voice: **Play the Hits, ONLY the Hits, ALL the time.**

Randy Michaels has two e-books published and available from Dan O'Day at

https://store.danoday.com/radio-e-books/

titled "Fundamentals of Radio Programming", and "Positioning Your Radio Station".

Much of what I have learned and trained over the years was seeded by Randy's enthusiastic and passionate presentation, which occurred over twenty-five years ago. Recently, I purchased these two books; it seems they are primarily based on Randy's presentation.

What a joy to read all that familiar material I had heard in Randy's original presentation, upon which, for decades, I had based my own training. If you never read anything else, these two e-books will set you on the right path.

Then there are people like Valerie Geller; I read and re-read all her books every year. Others are Mike McVay, Dan O'Day, Thomas Giger, Kurt Hanson of RAIN News, Francis Curry from the UK, Sean Ross who frequently has some fascinating insights on various playlist compositions. Mark Ramsey, too, often startles me out of my complacent frames of reference, as does James Cridland who continues to fill me with excitement and terms himself as a Radio Futurologist.

And many others, amongst whom is Phil Dowse from Australia, who, in 2005, wrote the book "The Number One Book on Radio Programming ... Ever!" It is a fine source of insights, tips, and tricks.

For over forty years, I have had the pleasure and privilege of working for RCS in Africa and the USA. For most of those years, the imminent "demise of Radio" has been frequently announced. That is not going to happen. Radio is here to stay forever.

Radio is free-to-air and available anywhere, at any time, including being a late-night companion in bed. For over 50% of the world's population, it is the primary source of entertainment, information, news, and education.

Over 95% of the population in much of Africa and South Africa hears the radio daily.

Even in the USA, I note that recently iHeart Media, with approximately 800 radio stations under their banner, proudly announced several percentage increases in their listenership numbers nationwide.

During my years with RCS, I have reviewed nearly 1,000 Selector and GSelector databases and trained hundreds more radio and TV music, talk, and programming personnel, as well as on-air personalities, in the implementation and use of these products.

Recently, I realized that on every one of those occasions, I had repeated basic philosophies and concepts each time. It was time to write this all down so I could invest more time using the products rather than discussing these philosophies and practices.

By implementing these philosophies, you will be empowered to maintain a consistent character for your station and deliver focused programming to enhance your listeners' experience.

Many of these philosophies are implemented in RCS's GSelector.

These philosophies also apply to most radio scheduling software. I am sure the imminent arrival of AI Programming scheduling software will also have at its core these same philosophies.

If you are new to GSelector, I strongly advise seeking training from one of your local RCS Support Engineers.

Once you feel you have mastered GSelector, it is crucial to continue seeking training. The capabilities available "under the hood" are truly amazing.

With their broadcasting background and real-world experience, the RCS Support Engineers can empower you with the nuances of programming with GSelector. This proactive approach will prepare you for the ever-evolving landscape of the broadcasting industry.

The future and the ever-changing landscape of options and competition are what we should all apply our minds to as the world of radio evolves at a breathtaking pace. If you are making a career in broadcasting, you must be obsessively fascinated and passionately involved in the industry.

This fascination and involvement will inspire and motivate you to excel in the broadcasting industry.

We should look to the giants in our industry for ideas and assistance. Please don't be ashamed: it's accepted practice in the radio industry to borrow their hard-earned ideas. That is precisely why they preach from the Pulpit of Good Radio Practices. If properly applied, they work.

An Overview

Let's start by describing the straightforward business of broadcasting.

We broadcast interesting, engaging, compelling, amusing, and relevant programming on our transmitter and on as many other media as possible. Let us not forget podcasts, internet streams, YouTube, and other highly relevant social media.

We gain listeners from a specific demographic, a community of like-minded individuals, and then sell access to that highly targeted community to marketers.

If we do well, our costs to broadcast the programming are far outpaced by the revenues we earn from advertising.

Broadcast attracts listeners, and advertisers want to pay for exposure to those listeners. Radio is a business where the incremental cash profit rises exponentially once your costs are met. We aren't like a widget manufacturer who, when his widgets gain popularity, must manufacture more and more widgets.

Our programming has a finite cost, and the increase in the price of doing business once that base-level cost is passed is minor.

Therefore, increasing our advertising revenue has a minimal impact on the cost of doing business, but a significant effect on the bottom line.

Radio can be hugely profitable if you get it right. But remember that even if you believe it is all about the "Art", investors keep score very simply. They look at the bottom line, the profit generated. Therefore, if we want to indulge in our passion, we must recognize that radio is a business.

So, what is this business all about? It is highly competitive. In Kampala or New York, approximately 30 stations fight for the same listeners and advertising dollars. So, how do you differentiate yourself from the pack?

The most important start is that you can be heard. Your transmitter footprint and the fidelity of your signal define your business's essential potential. The most engaging and fascinating program that your potential listeners cannot hear is not worth a dime. If they can't listen to it, it isn't very relevant.

Whenever I travel, I carry an excellent Sony portable radio; however, in major cities, it is often a fact that I can only hear about 50% to 60% of the available stations.

Address that issue so your station is widely heard, and you're well on your way to success.

If you are in the radio and programming business, you should be familiar with your signal. Get out there with several (cheap to expensive) portable radios and your car radio.

Travel around where you expect to be heard and even where you don’t expect to be heard. How does your signal perform against your competitors? How good is the fidelity of your signal? You don't have a business if you cannot be easily heard in high fidelity in all the right places.

Once you have established your transmitter footprint, you can investigate the potential audience and the programming that will attract and breed loyalty to your brand.

Then, of course, there are sales and the management of careful cost control. That puts a peculiar slant on our business hierarchy.

First – Engineering to ensure we can deliver a signal 24/7/365.

Second – Programming – our primary product.

Third – Sales and ensuring that we control our costs satisfactorily, so that sales exceed costs.

That's something that's intuitively easy to understand in radio, but it needs to be remembered amidst the flurry of the broadcast business.

Unlike Unilever, whose products dominate supermarket shelves, we only have two products and two clients.

The first product is our programming, which is directed at the listener, our primary client. The secondary product is the listener, the product for the advertiser.

What comes first: the chicken or the egg? In broadcasting, there is no puzzle. Programming comes first; Sales comes second. Without one, we don't have the other. Of course, without sales, we cannot indulge our passion.

I have heard stations in Lagos with 26 minutes of advertising per hour in their drive shows. In New York, I often wonder when I will be able to listen to any programming; it seems that a significant portion of the airtime is dedicated to commercials.

When I suggested that the station might alienate its listeners, if not aggressively chase them away, I was informed, *"The advertisers insist on buying that daypart."*

Broadcasters would do well to remember that the listener is our primary client. Not the Marketer.

And they should be reminded of the fundamental business principle of supply and demand, which dictates price.

With a finite number of avails per hour, if clients are all clamoring for the same limited supply, this presents an opportunity to price accordingly and earn significantly more money without harming the golden goose.

Remember this Clear Channel initiative many years ago? *"Less is More"* aimed to reduce the number of spots per hour while increasing the marketer's costs and the station's revenue. It was reportedly remarkably successful in its implementation and was copied by radio stations worldwide.

Programming comes first. Well, at least before sales and after engineering, which will ensure we have a decent signal to broadcast our program.

Of course, there is a constant tension between sales and programming.

However, we both share the same objective: to build a valuable audience demographic that can be sold to our marketers and generate a decent income for us and our shareholders. Beware the allure of the mighty dollar.

If the sales manager approaches the program manager with marvelous news:

"We have a client who will pay us a million dollars for a week of programming. They only want exclusive use of our 7:00 am – 8:00 am Morning Drive show to discuss recipes using their product."

The answer must always be 'No!'

Devoting an hour of your prime time to recipes can damage your hard-earned brand image and possibly lose your audience forever. We need to retain the million dollars, but we should utilize it differently: for sponsorships, promotions, prizes, and generating excitement. All this without destroying our brand.

Having that amount of dollars available enables you to claim and probably hold the title of "XYZ-FM – The station with the biggest prizes" alongside your other claim of "XYZ-FM – The station with the best music"

But how many program, general, and sales managers believe that? To be a successful PD, you must convince yourself that this is true.

So, is it Maths, Art, or Magic? Radio and TV music programming remains a mix of all three: it's maths, art, *and* pure magic.

Regarding mathematics, we now have access to excellent software tools that assist in researching and scheduling our programs.

The problem is that the tools are often not used effectively. And if you are solely dependent on math, that, too, can be a problem.

Yes, radio programming certainly has a considerable component in math.

I recently reviewed the U.S. Patent for GSelector and its associated scheduling algorithms. Shortly into the patent, I realized that I was out of my depth.

The mathematics in these algorithms is so incredibly complex that they reach what is scientifically known as "Chaotic Maths". The variables that determine the scheduling of a specific song in a particular position are indeed chaotic.

A review of the audit trail for how a song is ultimately scheduled in GSelector will confirm the complexity of this process.

But it’s more than math—much more.

It’s not about researching 1,000 songs and claiming the top 100 as your own. It's far more complicated.

It’s not just about which songs are on your playlist; it's about how the internal balance of your entire playlist is composed.

It’s not just about the specific music you play; it’s much more complex. It involves the sequence in which you play them, as well as when and where they are placed.

Your station is defined by the order in which you play the chosen songs. Play well-researched songs in an appropriate category, which determines how often they are played.

It is about where you position each category of song.

And it's about how you cradle newer songs you want to play alongside core, familiar, much-loved songs, so that they become familiar too.

It’s about being aware of what your competitors are doing.

It's about how you and your DJs identify with your listeners and ensure that your station is *"My Station, playing the soundtrack of my life."* It is also about how empathetic, amusing, compelling, informative, and revealing your on-air presenters are of their inner selves.

Our presenters are the key differentiators that set us apart from our competitors and other media, including streaming. They are what keep us relevant and engaging.

Competing with a station that relies on its PD's "golden ears" and "years of experience" without respecting the available math is an easy game. You can easily beat them with some research and a tightly formatted station playing hits.

Randy Michaels again: He related how he entered the Tampa market with Power Pig, competing against the then-market leader, Q105, which was the top station among teens, 35-year-old men, and women. Q105 and played a mix of AC, Rock, and everything. They lacked focus.

Power Pig came on air with only 90 tightly focused songs and 105 liners playing music that Q105 did not.

Within 74 days, the entire market matrix was disrupted, and Power Pig became the number one station in listeners and revenue. And it remained high in ratings for many years.

So, programming is mathematics, with a healthy dose of art tempered by your instincts and gut feel.

The tools are there; use them and respect them but apply your creativity too. You will be hard to beat.

What follows is not intended to serve as an instruction manual.

For that, I will direct you to RCS trainers, the excellent online in-context GSelector HELP, the wealth of videos, weekly RCS YouTube sessions, and the considerable library of documents and videos that RCS has built over the years. RCS also offers an excellent online academy that awards a worldwide-recognized certification upon completion.

This book primarily provides essential but vital philosophies and practices.

By implementing these philosophies, you will be empowered to maintain a consistent character for your station and deliver focused programming to enhance your listeners' experience.

The Program Director (PD) is a magical beast. One who stands responsible for the demands of the board, senior management, chief engineer, sales manager, finance manager, the audience, and those weird and wonderful people who are the engine room of a radio station: the engineers, scheduling, programming people, production people, producers, and on-air personnel.

The PD requires the political savvy and negotiation skills of Henry Kissinger, the brute strength of Atlas, and the empathy, compassion, and insights of an exceptional psychologist. The PD must act as a benevolent dictator, sometimes in a brutal manner and sometimes with sympathy and compassion.

The PD needs to be able to listen, comprehend the often hysterical and incomprehensible, and then lead this unruly bunch of type-A characters to commit to the station's vision.

One of the most important aspects of a PD is conducting frequent air-checks with each presenter, providing criticism, praise, and guidance in line with the vision, and ensuring they are building, rather than damaging, your brand.

The PD's vision for the station is to thoroughly define our potential audience, understand their needs, and identify our target marketers. How will we deliver to meet these diverse requirements?

The PD defines and curates the station's vision, upon which we build and consistently control a product that attracts a specific marketable listener demographic. And, particularly in countries where local content is mandated, ensure you retain your license by adhering to the terms of the license.

Being squeezed between the bean counters, engineering, and sales is a tough job, and there isn't a game plan out there that you can easily steal or acquire from a consultant.

I have seen and investigated over a thousand Selector databases in the past thirty to forty years. Each is different in some way, but the good ones—the high-income-earning stations—adhere to the same basic philosophies.

Be sure you are doing nothing on your radio station that you haven't thought through that makes sense to you in the context of your vision, the listeners, and the marketplace around you. Please don't do it because corporate tells you to, or because that is the formula presented by your consultant.

Following a consultant's formula without understanding and analyzing why the sequence occurs is lazy and probably wrong for your listeners.

We need to develop a vision, a formula, and a set of clocks populated with music and imaging that defines us and our listeners.

Who does this? Primarily, this responsibility falls to the Program Director (PD). Next, the PD needs to pay attention to the entire programming matrix.

This includes carefully researched music and imaging, the spots, where they occur, for how long, and the presenter's content.

Spots ought to be carefully reviewed for quality and content. Inappropriate spots for your specific audience can harm your brand, and poor technical quality is unacceptable.

How does the rest of the world interact with this? Our website, social media, blogs, podcasts, internet streams, and presenters all play a part.

This brings us to the age-old question: Who is the star? The Music or the Presenter? For someone like Howard Stern, the answer is simple. For the rest of us, it's more complicated. I suggest that the star is the music if it accounts for over 60% of your airtime. Build your vision around that fact. *After* you have read Brad March's comments on this issue:

Your presenters, the on-air talent, need firm direction, and you must pay attention to them.

Regardless of their stature in the market, all talent craves attention and needs ongoing criticism and praise.

The more attention you give them, the better your vision for the station will be followed through.

Note, too, that, without fail, the better revenue-earning stations treat their on-air logs as sacred; the on-air presenters must adhere to them, making no changes on a whim.

That vision needs to be informed by careful analysis of the marketplace, the history of your own and other stations in the market, and the potential listeners you want to attract.

You need to understand your competitors, what drives them in sales and programming, their goals, and their vision. Create your vision and execute it with enthusiasm and passion.

Music Rotations/Song Categories

First, a basic understanding of music rotation is necessary. In any music format, songs can be categorized into three basic types.

You can have more than three categories, but this is for illustration's sake. The categories refer to how often you want each song to be repeated on your station. Ultimately, every song in each category deserves equal plays to its peers.

These categories are not necessarily defined by genre (such as rock, pop, or R&B) or era. At this point in *my* life, my oldies are all from the 1960s, 1970s, and 1980s. For a teenager, their oldies might be from one month ago.

We must decide how often our listeners will hear the same song in each category.

Whatever you call the category is up to you, but conventionally, the first is 'Hot' or 'Current' - songs that are testing in your research, and most strongly fit the character of your station, ideally. How often do you play a hot song? Take a moment to examine your personal listening habits.

Let's say you hear of and download a new song from one of your favorite artists for your personal use.

How often do you play it during the day? Once, twice, or many times? The answer is that you will play it a lot. Five, ten, twenty times per day.

And you will continue to do so until the song is familiar, and you have learnt all the nuances of the lyrics and music.

Now apply that practice to your station. Once it becomes familiar, you might stop listening to that song for a while. Inevitably, if it deserves to be one of the best songs ideally suited to your audience, it enters your Core, Current, or Recurrent listening category at your station. You would now hear it once or twice per week. But, often, there is a danger that liking for the song would burn out.

Some songs that define your history and who you are, are moved to the third major category: Memory Makers or Oldies. You want to hear these songs occasionally; they bring back great memories for your listeners and help define your station.

For clarity, I have used only three broad categories in my explanation.

However, you can create many different categories and rotations to suit the specific character of your station and your listeners' demographics.

Here is a comparative table summarizing the typical names and rotation/repetition patterns for Adult Contemporary (AC) radio categories in Australia, England, France, Africa, and the USA, along with their standard playlist structures and rotation patterns. The rotation patterns are generalized, as exact numbers can vary by station, but they reflect industry norms.

Adult contemporary (AC) radio stations generally operate with moderate song repetition compared to formats like Top 40 or Contemporary Hit Radio (CHR), which feature much higher repetition rates. The frequency of song repetition in AC formats typically ranges from about 4 to 8 times per day for the most popular current songs, depending on the specific country and station's playlist strategy.

Typical Repetition	Playlist Size	Global Trends
For most adult contemporary stations, the heaviest-rotating current hits are played every 3 to 6 hours, resulting in 4 to 8 plays per day for top tracks.	AC stations usually rotate between 250 and 500 songs, which helps maintain a balance between familiarity and variety	While specific repetition rates can vary by market, the pattern of moderate repetition is consistent across major regions (USA, Europe, Australia), with France's "Variété FM" sometimes leaning toward less repetition and more variety.

Names for various Adult Contemporary formats

USA	England	France	Australia	Africa
The AC format is highly segmented: Hot AC (younger, up-tempo), Mainstream AC (broad appeal), Soft AC (mellow, female-targeted), Urban AC (R&B/soul), Rhythmic AC (dance, pop), and Modern AC (rock/pop-rock).	Often uses "Adult Contemporary," "Hot AC," or "Smooth AC," with "Easy Listening" sometimes used for softer variants.	"Variété FM" is the most common, referring to a broad AC format with French pop and international hits, sometimes grouped under "Adult Contemporary" or "Easy Listening".	Similar to the US and UK, with "Adult Contemporary" "Hot AC," and "Easy Listening" as common labels.	Similar to UK and USA.

Number of songs in a typical AC database

Typical AC	Hot AC	Mainstream/Soft	In Africa
250–500 songs, depending on the sub-format and station philosophy.	Tends to have a tighter playlist (250–350 songs) with more current hits and faster rotation.	Broader playlists (350–500 songs), including more classics and less repetition.	To buck the trend, regardless of format, stations have an active playlist of from 500-1,000 songs. There are some with over 3,000, or more. That does have a significant impact on rotations.

Typical Song Rotations

Mainstream/Soft	Hot	Variété FM
Mainstream/Soft AC: Songs typically repeat 4–6 times per day (every 4–6 hours)	Songs in heavy rotation may repeat 6–8 times per day (every 3–4 hours in a 24-hour cycle)	Common in France. Often emphasizes variety, with songs repeating less frequently (3–5 times per day)

Rotation Patterns

General Rule	Heavy Rotation	Medium/New Currents	International Consistency
The tighter the playlist, the higher the repetition; broader playlists mean less repetition.	**Refers to the most popular current songs, which are played more frequently, sometimes up to 8 times per day in Hot AC formats. A Top 40 station might repeat the same song as many as twelve (or even more) times per day**	**Newer or less popular songs may rotate less, sometimes only 2–4 times per day, or even less for "gold" or classic tracks.**	**The names and rotation patterns are broadly similar in Australia, England, and the USA, with France being notable for "Variété FM," which is a uniquely French term for a similar format. Africa goes its path with a massive variety of different category names.**

Daily rotations and the number of songs in the active playlist

Country, Generic format names, Number of songs in playlist, Number of times the same song repeats per day.				
USA	**England**	**France**	**Australia**	**Africa**
Hot AC, Mainstream AC, Soft AC, Urban AC, Rhythmic AC, Modern AC 250–350 (Hot AC: ~300; Mainstream/Soft: (350–500) 4–8 (Hot AC: up to 8 times per day Mainstream/Soft: 4–6 times per day	**Adult Contemporary, Smooth AC, Hot AC, Easy Listening (sometimes) 300–400 (similar to US) 4–6 times per day**	**Variété FM, Adult Contemp orary, Easy Listening 350–500 (Variété FM: often broader, classic-heavy) 3–5 times per day (Variété FM: less repetition , more variety)**	**Adult Contemporary, Hot AC, Easy Listening 300–400 4–7 times per day**	**Adult Contemporary, Hot AC, Easy listening 600-1,000 Hot 8-12 times per day Core/Currents 1 per day Oldies repeat every fifth day.**

Regardless of the song category, we must ensure that our station plays only the Best of the Best.

Remember that, in general, you have only approximately 250 slots available for songs per day. Don't waste any one position on anything but the best of the best.

Categories must be set to a defined number of songs and clock calls. This results in the required number of times a song naturally rotates and the frequency at which a listener hears the same song, which meets your vision.

How long should a song remain in a category? It's not about a specific number of plays for me. It is…for as long as it deserves to. Research will tell you when to move or rest a song.

There are iconic songs that have the gift of immortality. Led Zeppelin's "Stairway to Heaven" was released in 1971, while Dire Straits' "Sultans of Swing" and Queen's "We Will Rock You" were released in 1978.

Forty-five to fifty years later, these songs remain the bedrock of many adult contemporary or rock-oriented playlists, appearing on the station several times a week.

High song rotations and small playlists are hard to sell to even experienced programmers. They fight it every turn, claiming listeners who email, SMS, or phone in and complain that we are playing a specific song too often.

Perhaps those listeners would not call in and complain if they love the song you are playing "too often"?

They are more likely to mean that you are playing the wrong song too often.

It should be considered statistically insignificant if one or even a hundred listeners complain about a song being played too often.

However, the common belief is that there are far more people who will switch stations, rather than complain. So, if a thousand people complain, perhaps you need to review the song or spend more time listening to your station.

It is an age-old programming philosophy that you should pay homage to all the music types and genres that comprise the consistent character of your station and its listeners within a 15 to 20-minute set.

That is a big call, and only some may be possible, especially if you have a multitude of categories that you rate as necessary.

But a mix of hot, core, and memory makers in every set goes a long way toward satisfying the needs of your carefully researched but diverse listeners.

For this exercise, let's assume someone listens to your station for two hours a day.

Thus, we hope this specific listener pays attention to your station for one-twelfth (two hours) of any day.

We have several objectives during this specific listener's two hours of listening.

We want the listener to hear every one of our HOT songs several times, at least once every day, possibly more times per day.

We need that new HOT song to become familiar and embedded in our listeners' consciousness.

If this occurs in those listeners' two hours per day, a HOT song will have to be rotated through every two hours of the twenty-four hours. Thus, in this example, hot songs are repeated twelve times per day.

We want them to hear every core song at least once every five days. Why? Because it has been researched and is part of our audience's history and our station's character. It defines who they and we are.

And it follows their usual personal listening preferences.

For core songs, if we repeat the same song at least once per day, somewhere within the 24 hours, it is only likely that this listener will hear it every 12 days. Even this rotation might be too slow, considering one-twelfth of the day is spent listening.

During our two-hour listening session, we want our listeners to hear the same memory-maker song at least once every two to three weeks.

A memory-maker category with content of 200 songs that repeats somewhere during our day once every ten days is only likely to pass through our two-hour window every twenty days. Perhaps too seldom?

We require only a few active songs to deliver adequate rotations.

The largest revenue-earning stations in Africa and the rest of the world generally have an active playlist of fewer than 600 songs, some significantly fewer.

Remember, too, that you are, at any time, using a playlist that is a tiny subset of the entire music universe appropriate to your station. So, ensure that you review the whole playlist for every music meeting.

Replace songs when necessary and substitute them with appropriate alternatives from your extensive inactive library to keep your playlist fresh and engaging. Large categories should be updated regularly. This may mean changing up to 10% of your core songs every two weeks.

If we end up with a vision that requires eleven songs in the HOT category, that number must be cast in stone.

At a music meeting (which I hope—no, I am sure you all do at least once a week), several decisions must be made when a new song is considered for entry into the HOT (or any) category. If there is any doubt, the benevolent dictator, the program manager, should make the decision.

- Which of the existing songs in HOT will we sacrifice to enable the new song to enter the HOTs?
- What do we do with the sacrificed song? Dump it, rest it, or move to another category?
- If we move it to another category, which song will we sacrifice in that category, and what will we do with that song? Dump it, rest it, or move it?

When reviewing a database at a new station, one of my first questions is precisely what rotation is expected for the various song categories.

Almost without exception, their belief in what they want to achieve regarding their vision is not reflected in the true rotations of those songs.

They usually play each song far less often than expected. Frequently, the power plays are far too frequent for songs that have yet to be researched as deserving a place in this Power category.

Why is this happening? Why are PDs not effectively utilizing the tools available to them?

There are many reasons for this, but gut feeling and peer pressure are among the most significant factors.

Remember that any song follows a standard product life-cycle bell curve of:

- Introduction,
- Acceptance and growing familiarity,
- Maturity - Liked,
- Next, a decline occurs down the opposite side of the curve.

Often, our instinct tells us that a song has matured and is starting to fade, long before it even becomes familiar to our listeners.

A hot song needs to be heard often for it to become familiar.

If we are programming correctly, our listeners may hear it only once or twice a day. Even once or twice daily, the song might need many more plays to become familiar and loved.

The PDs and on-air talent hear the song many times per day, so it becomes familiar to them much quicker than to our listeners. Often, we want to dump a song well before it is tested as even "familiar." Far too frequently, we listen to our competitors and react to finding they no longer rotate a specific song as much as we do.

Alternatively, trade publications report that it has lost popularity. Or the record company may be trying to promote something else and suggest dropping specific songs.

We might think we need to let go of it as well, fearing that our peers will mock us for clinging to this song for too long.

Even worse, we might drop a song because our management insists that they've heard it enough.

None of these is a valid reason to discard a song. Neither your management nor your competitors should program your station. Your primary concern should be what is right for your listeners.

Another philosophy is that, regardless of format, we must play Power Core and Core/Current/Recurrent songs for at least 50-60% of our playlist. Why? Because these are the songs that define who we are, and who our listeners are.

These songs appeal to the broadest possible spectrum of our target audience. Remember, too, that having cross-over formatted songs and artists in your database is valuable. They can help you attract and retain listeners from competing stations.

Non-Music Events

GLinker handles the scheduling of all your non-song items, including live talk, pre-recorded features, station promos, program promos, competitions, promotions, news, traffic, and weather.

Scheduling LINKS ensures that the on-air presenters know precisely what is required every hour. The magic of Linker is that you can use almost all the same attributes as Songs, and program accordingly.

Every item that occurs in each hour needs to be detailed to enable the correct timing of the hour.

Note that the rundown graphic below includes everything scheduled for the hour, along with links, spots, music, and exact time placeholders. Each item also includes the expected broadcast time and runtime duration for that event.

Time	Item	Date/Performer	Duration
7:00:00	Exact Time Marker	13/4/07 7:00:00	
7:00:00	Top of the hour signal		0:05
7:00:05	News Jingle In		0:09
7:00:14	News 07:00		5:06
7:05:20	News Jingle Out		0:07
7:05:27	Coming up in the Hour - 07:05		0:31
7:05:58	Discussion - Arms Deal Bribe		3:59
7:09:37	Discussion - SA Rugby Prejudice		6:45
7:15:00	Exact Time Marker	13/4/07 7:15:00	
7:15:00	Spotset Holder		2:00
7:17:00	AM Live Jingle		0:03
7:17:03	Time Check 07:18		0:01
7:17:04	Discussion - Red Cross		6:02
7:23:06	Discussion - WHO Situation of Hos		5:00
7:28:06	Discussion - Rugby Politics		1:42
7:30:00	Exact Time Marker	13/4/07 7:30:00	
7:30:00	Traffic Mention Live		1:00
7:31:00	News 07:30		4:51
7:35:51	News Jingle Out		0:07
7:35:58	Traffic Mention Live		1:00
7:36:58	Traffic		0:43
7:37:41	Finance Mention Live		0:01
7:37:42	Finance News		1:52
7:39:34	Sports News		4:33
7:44:07	Coming Up - After Eight Debate		0:13
7:45:00	Exact Time Marker	13/4/07 7:45:00	
7:45:00	Spotset Holder		2:00
7:47:00	AM Live Jingle		0:03
7:47:03	Book of the Week		1:05
7:48:08	Discussion - Iraq Situation		5:59
7:54:07	Discussion - Maize Farming		4:46
7:55:00	Exact Time Marker	13/4/07 7:55:00	
7:55:00	Spotset Holder		1:00
7:56:00	Station Promo Jingle		0:07
7:56:07	Layla	Derek_and_the_Dominoes	1:52
8:00:00	Exact Time Marker	13/4/07 8:00:00	

This is from a Talk station. The pale-yellow events are LINK events, detailing what is expected at that time. The grey events are EXACT TIME MARKERS, detailing when a specific event is due to occur.

The green events detail where we hope to place commercial spot breaks, and the blue event details where we expect a song to play.

As it happens, only one song at the end of this hour allows for filling if the talk subject doesn't take hold and fizzle out or doesn't work.

Provided everything is detailed, the on-air presenters are then given an exact plan of what is expected and when it is scheduled to occur. This ensures consistent control of your station's flow, content, and character. You can then reliably report what happened on air for management and statutory purposes.

Also, by adding notes to each link (or music) event, detailed information can be made available to the Presenter when on air.

This further enhances the presenter's ability to deliver a carefully curated and informative program, thus enhancing the listener's experience.

Typical Link Categories		
Station Promos	Station Jingles	Program Promos
News	Weather	Features
Telephone In	Interview	Drama
Presenter Chat	Pre-recorded	Program Open/Close Themes
Time Checks	Competitions	Public Service Announcements
...and many more as required by your station. Note that, with the correct link categorization, you can carefully control when every event is scheduled or repeated.		

You need to develop a standard naming practice for links. This ensures that when you search for a specific link, it will appear in the correct alphabetical order.

For example: All "Presenter Chat" links could be named with the preface LIVE, such as "LIVE—presenter intro program events for this hour" or "LIVE—presenter back announce/forward announce."

Thus, when searching for program announcer events while building a clock, they will all be grouped alphabetically, starting with "LIVE."

I present some rules and practices. There are, in fact, no rules; you do what you believe is best for your vision. Nonetheless, it is a good idea to at least start with a firm set of rules to work with. Break them later when you know what you are doing.

A CLOCK must reflect everything that occurs on-air. This must include all songs, all Presenter live talks, all program themes, all pre-programmed inserts, all station jingles, promos, and spots. Everything.

On the subject of Spot block lengths. 3 to four three-minute blocks per hour is already too intrusive on your programming. Perhaps consider that the first and last spots in the break have most recall and value for the marketer. Therefore a position-in-block premium price is appropriate. And, perhaps only two-spot breaks is optimum?

Music Events

Programming music for a radio or TV station is a complex combination of math, art, and magic. It entails following basic rules and processes, relying on intuition, music knowledge, market research, audience analysis, and song research, while also addressing the needs of listeners and viewers.

However, at the core is a simple philosophy:

Your airtime is a precious resource. Do not waste even one minute of it.

It is vitally important to control the content and character of every item broadcast consistently

At best, you can only play 12 to 15 songs per hour. Thus, you have approximately 250 slots available for songs to be played within 24 hours, usually fewer due to busy daytime dayparts.

Therefore, we want to ensure that only the best songs are ever played on our station and that each song is rotated through every hour and part of the day, so that all our listeners have a chance to hear each of these carefully selected songs.

Rule: Only the best-of-the-best songs that meet your listeners' requirements deserve a position on your playlist.

Rule: All songs in each category deserve to be played as often as all the others in that category.

Rule: Every song in every category should play and rotate through every appropriate hour and daypart.

This is all about the fact that only some listeners (a very small percentage) listen to your station all day and night.

Instead, they listen at the same time each day for a short duration: morning and afternoon drive shows, late afternoon/early evening, or overnight when they cannot sleep and have no one to talk to.

These overnight listeners are among your most committed, engaged, and loyal audience members. They have little else to do. Please don't ignore them. They represent an extremely valuable demographic for salespeople to target with relevant marketing efforts.

We program only the best-of-the-best songs and ensure that every listener or viewer has a chance to hear every song. Thus, every song needs to be rotated through every hour or daypart consistently.

TIP:

Before you start adding songs to your playlist, answer the question:

Who is your listener? Does this song deserve a place on our station at this time?

TIP:

Each radio or TV station has two clients, each with a different product.

The first client is the Listener; their product is your Programming.

The second client is the Marketer; their product is the Listener.

Our programming should mimic our audience's regular listening habits.

This is precisely how we program a station: by selecting categories of music that define how often each song is to be played and repeated.

Research

Broadcasters should thoroughly research their music. We often don't because it is deemed "too expensive." But is it too expensive to do the job properly, beat your competitors, and earn much more?

But don't sweat the numbers entirely. Keep sane and also give credence to your intuition.

I heard this story on research from Randy Michaels. At the Country Radio Seminar around 1978, a panel of Industry experts discussed research, and a morning guy from Kentucky was on the panel. He listened to everyone, and when eventually he was asked to comment, he said:

"Well, that's the problem with y'all's stations. You do that research, and then you feel like that's what you've got to play. We can't afford Billboard; we don't know what's Number One. So, we avoid a lot of awful records that way.

When the music comes in, we sit down and listen to it — the engineer, salespeople, and the presenters. The good ones we put on the air, the crappy ones we throw in the garbage!"

Keep it simple; and rely on your instinct…but not entirely.

Not researching at all is a very tightrope-risky business.

Do your call-out research, which requires a full-time person, at a minimum. Other than that, I suggest trusting professional researchers. Find the money. It will be worth it.

Call-out research involves phoning a random sample of people, playing them hooks of music, and asking them to rate each from 1 to 5, ranging from "Never heard it" to "That is familiar" to "Love it" to "Bored with it" to "Hate it".

The audience is first - who they are, their needs and tastes, and how we can best meet those.

Then, where is a specific song's life cycle on your playlist? Are we playing it too much or not enough? Should a particular song be rested for a while or moved to a different category, thus achieving more or less frequent plays? Is the song "burning out" by being overplayed too often and breeding a dislike for it, or has it not even reached familiarity yet?

You also need to pay attention (at least some) to what your competitors are doing. What are they playing, and when and how often? Now, the matrix becomes complex.

Then there is auditorium research, where we gather a random sample of listeners and play up to 1,000 songs, asking for comments on each song.

Certainly, call-out and auditorium research have some value, but both methods require improvement. First, you should only empanel individuals who are your listeners or potential listeners.

Why would I want a hard-core heavy metal fan to comment on my adult contemporary playlist? Why am I interested in a 65-year-old grandmother's opinion when my target audience is 15-25?

You want to be sure you are researching people who are at least potential listeners to your station.

Sometimes, we want to specifically target our P1 primary core listeners who are loyal to our brand. They are the individuals we need to monitor closely regarding where a song sits on the product life cycle bell curve. Observe them intently; consider what troubles them, excites them, and pleases them. For this type of research, we require very different tools, some of which are available through RCS.

When I look at logs and analyses, I often see anomalies resulting from research figures. Be careful. Do research, but don't be a slave to those numbers.

Remember that a song testing at seventy-five could be in the adoption phase, not yet familiar, or in the dying-swan falling-off-the-curve phase.

Beware of research figures that negate everything you thought to be true.

Your gut feeling might be accurate, and the research numbers could be distorted due to poorly constructed questions or faulty calculations. Review your research and remain loyal to your vision and instincts.

After analyzing the research with your intuition, adjust the clocks and category content to honor it. Additionally, feature the Best of the Best more frequently.

In many countries, listener research is conducted through diaries. Indeed, the listeners don't record that they switched from Station A to Station B at 08:05:16. What happens is that they rely on faulty memory and log their listening starting from the beginning of an hour or half past the hour, then ending in the next half hour.

Previously, we thought certain portions of an hour were listened to more than others. As a result, we planned to attract and retain listeners during those times. Nielsen's PPM has debunked that theory.

There is no preferred time during the hour for programming. All times may hold equal value for the listener, but keeping the listener in to the next quarter hour should always be the golden rule.

Additionally, PPM has largely debunked the idea that spot blocks result in significant listener attrition. PPM has not proven this. At worst, PPM indicates that the number of listeners experiences a slight decline during spot blocks.

PPM cannot indicate whether we experienced a catastrophic loss during the block, but it is possible that we gained many listeners as they switched from a different station to ours. However, playing an extra juicy best-of-the-best Opener crossover song right after the spot block will significantly help retain any competitor's listeners who changed stations during their spot break.

Diary entries are usually filled in well after any listening occurred. How do they remember which station they switched from or to? The truth is they don't know.

They do remember that your station is loud and proud in terms of imaging at every opportunity, and thus, they distinctly remember switching to your station.

That is why you hear the most successful stations worldwide often playing station imaging between every song.

Commercial Research Sources

Several excellent commercial services are available that research different aspects of broadcasting worldwide. RAJAR, Edison, IPSOS, Kantar, Nielsen, and various audience listenership measurement bodies around the World and in Africa, such as the South African Broadcast Research Council, provide multiple forms of listenership measurements. The methodology varies.

Of course, if, according to published research, your audience has grown substantially over the most recent research cycle, you would likely believe this to be entirely accurate. However, if it has fallen, we consider that data to be inaccurate. But these services supply the only currency we have to work with. It always surprises me when I discover that many of our clients do not review audience research reports immediately after they become available. Shame on them.

Call-out Research

Then there is "Call-Out." We assemble five to ten hooks from songs we need to research, and phone random or specific people to ask them to participate in the research exercise.

The most significant problem with call-outs is that we must ensure we are speaking to existing or potential listeners.

It would be a pointless exercise to play ten Top 40 songs to an 85-year-old retiree and expect valuable insights to be offered.

For several years, one of RCS's clients launched a heavily repetitive and well-promoted campaign inviting listeners to "Join our Music Consultant Team."

Respondents signed on to the music consultant team on a dedicated micro-website. Now, our users can be tracked and can complete the research project at their leisure.

The station now knew they were talking to their existing audience, so the song's life-cycle information would be valid.

“Join our music consultant team” was very successful because it delivered deep insights into who our audience was and where a song was on the product life cycle curve.

Additionally, sponsorship could be sold, thereby ensuring that this was a revenue-generating exercise rather than a purely cost-driven one.

Requiring your sponsor to supply a substantial prize to participants, such as a high-quality cell phone every month, certainly drives people to respond.

You can also track a specific listener’s song preferences over time to see how their music tastes align with your vision. Ensure that you structure the survey so that you ask and receive answers to relevant issues.

The objective might be to ensure that your most loyal (P1) listeners are happy and being catered to.

A different objective might be to research potential listeners and establish who they are, as well as what is required in our music to convert them into P1 listeners. Ensure that your respondents are at least occasional listeners to your station, with a balanced representation of males and females, and within the correct age and income bracket you are targeting.

Alternatively, we may be trying to identify our listeners' demographic and social background.

This is particularly interesting in South Africa, where spot time is purchased based on the Living Standard Measure (LSM)—what kind of appliances do you have in your home, and where do you live? What education level have you achieved? What is your salary? What is your banking use? What kind of transport do you use? And lots more.

LSM 1-4 have the least access to wealth, LSM 5 -7 are “middle-class,” and LSM 8-10 have the most access to wealth, high educational standards, full access to banking services, participation in various leisure activities, and broad access to information. Each LSM group appeals to a different set of marketers.

For community and commercial stations, remember that we are broadcasting for and giving a voice to a community of like-minded people. Researching your mix of programming can be a challenge because each listener has a specific agenda to sell. Be careful.

Asking the wrong questions leads you to an inappropriate skew in information. What is the right question?

That is precisely why research companies are seen as costly. They understand what questions to ask and how to measure, interpret, and allocate the responses.

At least fifty to a hundred responses are required to calculate valid statistical values. Two hundred is better. Even more? Perfect.

Callouts should be repeated every two weeks, and you need to research P1 listeners from the past seven days. P1, your primary loyal listeners, must be examined to ensure you respect the life cycles of all your songs, neither underplaying nor overplaying.

For P2 and occasional listeners, many more questions must be answered, not the least of which is, "What do we need to do to attract them to become P1 listeners?"

And then there are those listeners who never listen to your station but would if you persuaded them to at least try it out.

You need to find answers to the most critical questions, especially for your P1 listeners: Who are they? What are their preferences in programming and music?

Researching each of these groups is a very different exercise. Be skeptical of all results, but don't discard them entirely.

- All the answers need to be treated as possibly suspect.
- Burn levels may be premature; not everyone is a P1 listener, so they might not have heard the song as often as required to become even familiar.
- Be tolerant of anomalies and unusual spikes in the numbers.
- Be prepared to discard off-the-chart answers and respondents.
- Accept that not all listeners fulfil your time-spent listening expectations.
- Spending more than six to seven minutes on a survey will quickly introduce survey resistance and paralysis.
- Repeatedly calling the same person too often will also do that.

- "Do you listen to XYZ-FM?" if closely followed by "How long do you listen to XYZ-FM each day?" and "When do you listen to XYZ-FM?" will quite possibly elicit a response that they believe will please you. So, ask this in different ways, possibly at least twice in the interview.
- Where do you listen, how often, and when? These are all questions that must be carefully crafted and analyzed.

Tabulate this data to make sense to you and then review and question it. A lot.

The addendum at the end includes portions of a typical call-out research project. Additionally, a "perceptual" research project analysis aims to gain a granular understanding of a station's audience. Both projects used RCS's "Test all Media©."

In this case, these were not true "callout" research projects; they utilized a micro-website attached to the station's primary website for participants to log on to and fill in their answers.

This allowed the participants to complete each project at their leisure and to be tracked continuously over more than a year.

Hence, the data is very granular and rich.

Additionally, it allowed the station to track specific users' likes and dislikes over this period, providing them with profound insights into who their listeners were, what each person's particular tastes were, how quickly a song resonated with them, and much more. Indeed, conduct extensive research, always.

However! Treat all research with skepticism and care. Brad March warns:

Too often, research is misused to justify poor decisions, and when it's in the wrong hands, it can lead to disastrous consequences. This is particularly true for talent research. I've witnessed numerous TV personalities who test well, fail miserably on radio. Just because research indicates a strong (TV) profile and likability, it doesn't mean that the same talent will resonate on radio or other media. You cannot rely solely on this data.

Off-Air Monitoring Research

One of the most comprehensive and value-for-money data sources worldwide is RCS's Mediabase and Media Monitors, branded in some countries as Aircheck. https://www.mediamonitors.com/

https://www2.mediabase.com/mbapp/Account/Login

These services provide valuable insights into radio and television advertising, music played across various regional stations, and other comparable sales data across the market. Who is advertising at a competitor's station and not at yours? Why? And lots more insights are available to you, both in terms of the music you are playing vs. your competitor and in terms of product market research.

In Africa, Aircheck (formerly RCS's Media Monitors, now rebranded as Aircheck) is franchised to Afstereo.

In their airplay charts, additional information is available, including the number of Spotify streams the song received, the number of YouTube views it garnered, and more.

The Afstereo data is accurate, virtually real-time, and genuine.

For more information about Off-Air Monitoring Research companies, check out these URLs:

- Afstereo https://www.afstereo.com
- Media Monitors https://www.mediamonitors.com/
- Mediabase https://www2.mediabase.com/mbapp/Account/Login
- IPSOS https://www.ipsos.com Established in France in 1975, before expanding to the rest of the world in 1990, the IPSOS positioning statement is worthy of remembering: **YOU ACT BETTER WHEN YOU ARE SURE.**
- Broadcast Research Council of South Africa https://www.brcsa.org.za This covers TV and radio audience numbers, and the information is free.
- Nielsen https://www.nielsen.com In the USA, their Portable People Meter (PPM) revolutionized the previous method of conducting audience diary research, debunking many long-held programming beliefs and casting doubt on established practices. There remain objections to the validity of PPM, but this is indeed progress.

Social Media

Social media has become integral to our broadcasting matrix, serving as an excellent source of market and research data. We must be inventive and explore the full potential of social media in our work.

For instance, I am aware of one international group of Top 40 stations that relies almost entirely on Billboard and YouTube for music research.

Billboard's Top 40, trending songs, current hits, and recurrent tracks are all meticulously tracked on YouTube. The number of views and an analysis of their trajectory are central to their programming philosophy. Rising, falling, and stagnant interest influence how they categorize and schedule.

There are also many free new music services like AllAccess (https://www.allaccess.com/) and NewMusicServer https://www.newmusicserver.com/, which allow radio stations to break new top 40 releases and other genres.

The joy of both these services and several others is that they are free, and you can download high-fidelity .wav files rather than heavily compressed and fidelity compromised .mp3 files.

In the digital age, radio broadcasting is no longer confined to the airwaves. Social media has revolutionized how radio broadcasters engage with their audience, promote content, and expand their reach. Effectively using social media can transform a traditional radio station into a dynamic, interactive platform that resonates with today's listeners.

Social media allows radio to return to its former glory: a community of like-minded people discussing and experiencing relevant issues that speak to their minds and hearts.

Beware that when your on-air personnel use social media and refer to it on-air, they do so only on your station's account. We are in the business of building our station's brand, not the presenter's personal brand.

The first step in leveraging social media for radio broadcasting is understanding the audience. Different demographics prefer different social media platforms.

Younger audiences might be more active on Instagram, Snapchat and TikTok, while older listeners might favor Facebook, X, LinkedIn, and Bluesky.

Conducting audience research can help broadcasters identify where their listeners spend their time online.

This knowledge allows broadcasters to tailor their social media strategy to meet their audience's preferences and behavior. “Content is King” is on social media, as it is on radio. Radio broadcasters must create content that promotes their shows, builds the station’s brand, and engages their audience.

Social media should be as much a part of your station’s character as the actual broadcast. Strawberries AND cream, not Strawberries OR cream.

Your social media posts and presence should be about building a community of like-minded people. We all crave recognition and interaction with people who embrace our values and are like us.

Remember that radio was the original community of like-minded people well before the arrival of digital social media. Radio enabled people to have a voice and to argue their points.

It brought them together with people they could relate to and empathize with. This community appeals to similar groups and provides entertainment, engagement, amusement and information.

Share videos showing what happens behind the scenes at the radio station. This can include snippets of live broadcasts, interviews with hosts and guests, or studio tours.

Despite the proliferation of live-streamed radio, I am not convinced that video streaming from a live radio studio offers compelling programming. Radio is the theatre of the mind, and that theatre suffers disappointment when that incredibly sexy young voice that seduces and captivates you is from a seventy-five-year-old. Seldom does a radio studio offer gripping video scenes.

Social media enables us to add additional value to our broadcast. Extended interviews, more elaborate explanations of the issues, and references to other websites and sources are all excellent additions to our programming. Tease these interviews on social media to drive traffic to the radio broadcast. Equally, tease the social media on air.

Engage the audience by asking questions or conducting polls related to show topics.

This encourages interaction and provides valuable feedback, encouraging our listeners to contribute to the community discussion.

Encourage listeners to share their stories, photos, or videos related to the show. Featuring user-generated content can foster a sense of community and loyalty.

Social media is especially beneficial during talk shows. In-depth information and alternative views can be presented in real time. And don't forget to drive people to your website for more details on the subject under discussion.

Regular sponsored events should occur and be well-covered on air and on social media. Club events, sponsored music events, annual sporting events, and visits to unique places should all form part of your arsenal. Competitions and promos that excite your audience are a definite way of breeding brand ambassadors.

These can use social media's power to draw, engage, and inform. One of our clients in South Africa hosts a three-day, two-night luxury cruise along the coast every year. It is fully booked yearly and a significant event worth being loud and proud about.

They feature several broadcasts from the liner at sea, mingling of the station's hosts and their audience, continual partying, and extensive sponsorship opportunities, so much so that this trip is a major contributor to the station's annual bottom-line profit as they also earn a lot of money from sponsorships during the cruise.

Engagement is crucial for building a loyal listener base, and social media allows for direct interaction with the audience.

This points to the reality that every station needs a full-time Social Media specialist—an invaluable addition to your team.

And let us not forget that your station website requires constant review, changes, and additions - DAILY.

Many station websites have remained unchanged for several months. Again, shame on the guilty stations.

When updating the station website, track the live discussion subjects with extensive information and commentary, and add in-depth discussions of news items whenever possible. Enhance the music played with details about the songs and the artist, and keep your community involved.

- Respond to comments and messages: never ignore your listeners' posts.
- Respond to all listeners' comments and messages. This confirms that your station values their input, and fosters a sense of connection and community.
- Regularly host Q&A sessions where listeners can ask the hosts or special guests questions. This interactive approach can make listeners feel more involved.
- Acknowledge listeners on-air; ongoing social media interactions on-air can bridge the digital and broadcast realms. Listeners will appreciate the recognition and be more likely to engage online.

Research analytics are vital for establishing whether your direction and intent are being met. All social media platforms offer some sort of post-performance analysis, some more extensive than others. Use them.

You will gain insights into the audience demographics and how often you are gaining attention. Now, refine your efforts based on these real-world analytics rather than wishful thinking.

Paid social media advertising can amplify reach and increase the impact of a radio station's messaging, and social media channels like Facebook, Instagram, TikTok and X can provide you with targeted advertising.

Now, a radio station can advertise to people in a specific geographic area, age group, or those with topics of mutual interest. Social media advertising can amplify shows, events, or contests, thus increasing listenership.

At the heart of using social media lies the need to build a community around the radio station, which can enhance listener loyalty and engagement. Social media is ideal for creating and nurturing this community of like-minded people.

Social media is not “just another medium for us to appear on.” Integrating social media with on-air content can create an embracing experience for listeners.

Radio hosts can conduct social media polls or ask questions during the broadcast and encourage listeners to participate online.

Similarly, social media can extend the conversation from on-air segments and provide additional content and engagement opportunities.

This is especially true of phone or in-studio discussions where additional insights and opinions can be shared in real-time on social media and your website.

This dramatically increases the perceived value of what you are discussing and allows for actual participation in the discussion by your audience.

Set up and stick to a daily schedule for each program to broaden your reach beyond the transmitter's footprint to all the other available options. Be regular and meet expectations. This adds value to your brand, turns listeners into brand ambassadors, and fosters loyalty and a primary audience (P1) listeners.

Again, Brad March has firm views on Social Media;

Radio must now contend with a variety of platforms, including social media channels like TikTok and Instagram, as well as streaming services like Spotify and Apple Music, which offer consumers convenient, often ad-free music options. While podcasts may still be behind in revenue generation, they're also vying for listener attention. Most podcasts revolve around either specific topics or are driven by personalities, but being a star alone isn't enough; the content must be engaging, relevant, compelling, and fresh.

The same rules apply. It’s likely we will see more crossover between podcasters and radio, similar to what has been done by the We Mean Well team and the Life Uncut podcast duo, Brittney Hockley and Laura Byrne, both of whom are finding success on radio. Podcasters like Toni Lodge and Ryan Jon have emerged as global superstars, establishing their brand "Toni and Ryan" by producing content that is relatable, entertaining, and unique, targeting a specific market segment. Importantly, their content is compelling, and they really know their audience well. Their success is evident, having amassed over 100 million downloads and gaining popularity not just in Australia but also in the US, UK, and beyond. They've skillfully utilized social media to promote their daily podcast while also enhancing their brand and entertaining their audience.

Right now, it seems that podcasters are making better use of social media than radio talent and networks.

Competitions

Telephone calls and competitions are very useful mediums for enabling your listeners to participate in their favorite station.

The sense of belonging is vital, and one of the most valuable occasions for listeners to participate in your programming is when you announce them as a winner, with them live on air. This gives them a chance to vent their delight, to tell all their friends to listen, and creates a memorable experience that enhances their connection with the station for all your listeners.

Offering a prize of a station logo-branded coffee cup or a pizza to the fifth caller should never be considered. It is insulting to both your listener and your brand, and offering such a trivial prize diminishes your brand.

In mid-2024, Sean Ross, in his weekly Ross On Radio newsletter (which you should subscribe to), mentioned some significant competitions primarily because they were offering substantial prizes. Such a prize says an enormous amount about your brand and its listeners.

He wrote: *It's mid-winter in Melbourne (Australia) now, and currently, something is going on with almost every major outlet, even though most are relatively straight-ahead:*

CHR Fox 101.9 has a "$5-Million Match" money song contest and a contest where two contestants will give their junker car to be crushed live at an event, with one earning $50,000.

AC Smooth 91.5 is giving away a trip to the Cook Islands. News/Talk 3AW is giving away a 10-day trip to China. Both are "tell us why you want to go" website entries. Heritage rocker Triple-M is sending a Listener on a "Rock the Boat" cruise with Australian bands as part of its "G.O.A.T. Song Countdown," but also doing a $50,000 no-repeat workday.

CHR Nova 100 is doing "Pay Your Bills." In a throwback to our pretexting days, bills are all submitted via the website, and names are called on-air.

Nova 100 has also made local headlines for winning at least the initial round of the breakfast battle against KIIS FM's now-syndicated Kyle & Jackie O show, with the local show they replaced.

Kyle & Jackie O are running a "Split or Steal" contest on KIIS FM in Melbourne and Sydney. "Split or Steal" is one of the psychology-based games that became popular in the mid-'00s, but I encounter less these days.

Promotions and competitions are about increasing cumulative listening time and building our Brand. Add to that the potential for sponsorship and for your audience to participate in and enjoy the opportunity to win, and it is indeed a win-win situation.

If you have a competition, let it be for a substantial prize. How substantial?

How about: *Exclusive Limos will collect you and your date from your home. You will then have pre-dinner drinks at the upmarket Coro's Lounge with our Star Morning Drive Show DJ Big John, and then a romantic meal at La Perle five-star restaurant. After dinner, we will chauffeur you in style to attend a few sets at the Comedy Club, and then you can go off to the honeymoon suite at the Hilton Hotel. After breakfast, you will return to your home in an Exclusive Limo.*

That prize is worth giving away, as it discusses your listeners' aspirations and your station's brand.

It also allows you to sell to multiple sponsors participating in this competition.

Give value to your sponsors. If a sponsor offers a significant prize, ensure your “trade exchange” meets or exceeds that exchange's rate-card value.

However, beware of competition burnout. Too often, and the energy generated diminishes. Too few, and you lose opportunities for engagement.

Streamline the entry process through a dedicated microsite or social media platform. Make it easy for listeners to get involved. Be sure to publish the exact terms for the competition on your website. That avoids upset participants and lawsuits. Your website is the ideal platform to outline the precise details of any promotion or contest.

Any competition requires extensive support and marketing, both on-air and online, including on your website and social media. Billboards and newsprint are also options.

Every component is vital. Use on-air mentions, social media, and other channels to spread the word and excite your audience.

The competition needs to become an “event” in the lives of your audience.

It isn't just the prize that counts; it is the ability of your audience to participate and become part of the event that needs a boost.

Every possible media and tool should be utilized at the event, including on-air broadcasts, social media interactions, website hints, and live appearances. All contribute to the momentum that builds to a crescendo when the winner is revealed.

The magic happens when you transform competition winners into brand ambassadors for your station and your sponsors.

Celebrate your winners on-air, on social media, and through any other channels that will amplify their stories and inspire others to participate in the future.

There is another standard belief that there is nothing new in radio. This may well be true; someone might have used every possible competition idea somewhere. This is no great disaster; it allows you to use anyone else's proven ideas as well. Spin it uniquely, and the idea will become yours - temporarily.

For a final word on competitions and events, I quote the highly respected Brad March and Scott Muller SCA's Head of Content strategy and Insights from an article in Australia's Radio Today:

Radio promotions put the wow factor into radio. Promotions, stunts, and contests can create word of mouth, increase awareness and recall, drive cume and market share, create compelling listening, and are great revenue drivers for advertisers.

Another, often unspoken but major benefit, is that a great, well-executed, multifaceted promotion can give a station a sharp point of focus and direction. And stations with limited marketing budgets can benefit from strong PR via promotions & stunts. As well as appealing to the minority of active contest participants, promotions should also appeal to the majority, the passive listeners.

Whereas contests used to be targeted primarily to ratings diary keepers, now they are an integral part of a station's brand DNA.

Australian FM radio is recognised as being at the forefront of executing brilliant, original, and creative promotional campaigns. Many Australian promotions have found their way onto US, UK, European, Asian, and African radio over the years.

Station Imaging & Program Promos

Promotions and competitions are designed to increase cumulative listening time and establish our brand. Don't expect your audience to know your schedule. Tell them. Often. What can they hear and when? Point them toward favored programs and highlights. Make appointments for them to join you.

But beware of overextended station or program imaging. Thirty seconds is OK; fifteen seconds is better, and seven seconds is best. Anything longer than 30 seconds consumes too much valuable programming and song time.

Imaging and promos almost always last longer because you allow your production people latitude to demonstrate their considerable experience and creativity. Imaging and promos should sell the station and its programs, not illustrate how fantastic your production team is at assembling, editing, and mixing a tour-de-force production element.

For station imaging, you need to promote the essence of who you are.

Aspirational phrases often tend to be way off the mark.

Running an imaging spot that states" XYZ-FM - We play all the latest hits" when you never break hits at all will not win you loyal hit-seeking audiences.

Be cautious of punting too many messages on various issues. You cannot be everything to everyone; many different aspirational claims for your station can confuse the listeners.

Station imaging is too often left to the on-air talent. They are supposed to throw it in from time to time from hotkeys. Don't do that. Schedule it in your log. That way, you have control and can ensure it happens.

The only items not needing to be on the log are beds and the on-air talent's specific bings, boings, explosions, and other punctuation sound effects. Off-air recorded snippets of phone conversations belong on hotkeys or stacks. Beware, these contribute to a change of the hour duration in the log.

Schedule it in the log if you want talent to talk and define a projected talk duration. Check the log after the broadcast to ensure they didn't speak for three minutes instead of one minute and skip one of your carefully crafted scheduled songs.

Don't expect the talent to remember to play program promos.

Every hour ought to have a complement of one, two, or more scheduled-in-the-log promos for the next quarter hour, the next hour, another program somewhere else in the day or week, and possibly one for the same program tomorrow.

Brad March, a legendary Australian PD, launched B105 with street vehicles called Black Thunders. This was a large fleet of heavily branded (and fully sponsored) black SUVs that spent the entire day roaming the city, handing out station-branded T-shirts and gifts, and reporting live as often as possible. When you saw one of them or a fleet of them approaching you, on the way to an OB, you *knew* B105 was on the scene.

Positioning your station in the radio battlefield.

There's a book titled *Positioning: The Battle For Your Mind* by Al Ries & Jack Trout , which is a bible about positioning. Certainly, worth a read. Herewith the key concepts:

At its core, *positioning* is about how a brand (or radio station, product, or person) is perceived in the minds of consumers—not what the company says about itself, but what the audience believes. The book was groundbreaking in shifting focus away from product features and benefits and toward perception as the battlefield. The central idea is that people are bombarded with messages, and only a few stick. To succeed, you must claim a spot in the prospect's mind and hold it tightly.

Ries and Trout introduce the metaphor of the "ladder" in the consumer's mind—each product category has its ladder, and the mind assigns brands to rungs based on perceived leadership and relevance.

For example, in the soft drink category, Coca-Cola is typically at the top, followed by Pepsi, and so on. If you're not already on a ladder, you need to find a new one—or a new category altogether—to own.

One of the book's key arguments is that it's better to be first than to be better. The first brand to claim a position in the mind (e.g., Hertz for car rentals, Kleenex for tissues) has a massive advantage because people tend to remember the first and forget the rest. In radio, the first station to claim a distinct format—"the classic rock station" or "all-news all the time"—often owns that identity, even if competitors later do it better.

If you're not first in a crowded category, don't fight it head-on—reframe the field. This is what Ries and Trout call "creating a new ladder." Instead of being the second soft drink, be the first diet soft drink. Instead of being the second Top 40 station, be the first station for "intelligent pop for adults." In radio, this might mean repositioning a station around a specific mood, lifestyle, or time of day, rather than relying on broad genre labels.

A powerful (and often underused) tactic is *repositioning the competition.* This doesn't mean a direct attack but instead shifting the audience's perception of a competitor.

For example, Avis's campaign "We're number two. We try harder" cleverly repositioned Hertz as complacent. In radio, this could be framing a rival as predictable, out of touch, or too broad, while your station is fresh, local, or "tuned to today."

Perhaps one of the finest examples I know of repositioning the competitors occurred in 1990, when Brad March, one of the best PDs Australian radio has ever been blessed with, launched B105 in Brisbane. Brad's brilliant positioning was to place B105 as "Nothing but Great Rock and Rock" and claiming to be the best and only rock and roll position on the dial. He did indeed play some rock; Led Zeppelin and the Stones, for example. But he also played Celine Dion and other less-than-rock ballad hits. B105's imaging subtly claimed that every other station was inferior.

Having claimed this, the implication was that whatever everyone else was doing, it wasn't "Nothing but Great Rock and Roll." Was this true?

Well, not really. But the listener's perception was Yes, because B105's imaging said so, and no one else did.

Thus, stations playing soft AC Hits, and rock were second or even lower down the totem pole. Suddenly, every other station was competed with a station being loud and proud about being *the* Rock and Roll station. In their first research cycle, B105 was ranked #1 among listeners. In total, B105 held the # 1 position for 16 consecutive years, interrupted only by one month in 1994. For some periods they reached a 35% share!

The mind hates complexity. The most successful brands (and media properties) use simple, consistent messages that can be grasped quickly. Think of slogans like "The ultimate driving machine" (BMW) or "15 minutes could save you 15%" (GEICO). For radio, this means precise imaging, consistent voice, and format discipline—trying to cover too many genres or identities muddles perception and weakens loyalty.

Beware of *line extension*—when a successful brand tries to stretch its identity across too many products or formats. This confuses the consumer and dilutes the core position.

If a rock station starts adding pop or dance hits, it risks alienating its rock audience and compromising its identity. Strength comes from focus. A radio station must know precisely who it serves and why it matters to them. Attempting to be "for everyone" is seen as marketing suicide.

Advertising's job isn't to create a position, but to *reinforce* or *communicate* it. Great ads don't sell the product—they sell the position.

This means consistent messaging over time that helps consumers associate a simple, memorable idea with your brand or station. Advertising that tries to be clever or wide-ranging often fails because it blurs the position.

Just like brands, people need to define what they stand for in the marketplace.

If you're a radio host, are you the funny one, the music expert, the community connector, or the late-night confidante? You need to own that idea in the listener's mind.

Positioning is not about what you do to your product—it's what you do to the *mind* of the prospect.

You either plant a new idea, reinforce an existing one, or change and diminish the listener's perception of a competitor. And once you're in the mind, your most important task is to stay there, with clarity, focus, and consistency.

For radio professionals, *Positioning* is a strategic task. Format, branding, imaging, talent, and promotions must all reinforce one simple, clear position. In a cluttered market, the station that stands for something specific—and reinforces it relentlessly—is the one that gets remembered and tuned in.

Here's a fresh, fully color-coded mind map focusing on the core functions of radio station positioning in a competitive market.

These are the core issues you need to address in your Station positioning and imaging, as well as in repositioning your competitors against you.

When engaged in selling radio, do to ignore the fact that one of your prime positioning issues is to insure thar radio (ALL radio, including your competitors) is seen as the only always available free to air media available to your marketer with the most extensive and least expensive reach.

News

I was initially shocked to hear a statement from a leading USA station's news manager: *"If it happened beyond our station's transmission footprint, it didn't happen."* After further discussion, it made sense. Radio news is primarily a medium for 'breaking' news. News of utmost concern to our listeners is anything directly impacting their community, families, income, or businesses.

Events beyond the station's transmitter footprint can occasionally affect our listeners or be of immediate concern to them. For instance, the BBC's coverage of the attempted assassination of Mr. Trump in July 2024, which they reported only three minutes after it occurred, was of interest to their listeners despite the event happening well beyond their transmitter footprints.

But the prime purpose of the radio in terms of news remains "breaking news." We are generally not in-depth investigators of news events. To some extent, this might also occur in more extensive news-oriented discussions or editorials.

Your web page should cover all relevant events pervasively and in-depth if you do your job correctly.

Additionally, if you are hosting a talk show about a popular or contentious subject, you should provide extensive coverage and references on our website.

You should repeatedly direct your listeners to this nuanced and ever-changing source of information. Yet another source of advertising revenue is sponsorship of your website's extensive information.

For breaking news, we need to gauge its importance to our listeners. A bus falling off a cliff in Argentina that results in considerable loss of life is, while horrific, of little genuine interest to a listener in Munich—unless, of course, there is a large Argentinian expatriate population in Munich. However, extensive flooding in Philadelphia is of considerable interest to our Philadelphia listeners, even if it is on the opposite edge of the city.

Much of our news broadcasts in Africa are based on a public broadcaster model. They are extensive, often exceeding ten minutes or more in a drive show, and include lengthy and tedious monologues from politicians or celebrities.

In the USA, news broadcasts typically last only one minute, often include audio clips (after all, we are a "sound medium"), and cover up to six events—your choice.

I believe any news broadcast of more than two minutes on a music station is too long.

If we are a breaking news source, we can provide brief coverage of each event. However, we can and should delve deeper into each event, offering more coverage on our website and social media. We should also ensure that a short audio clip accompanies our news event as often as possible.

Humans are creatures of habit. Suppose we expect the 8:30 am news bulletin, which is promised and forward sold to occur at 8:30 am. In that case, it should occur strictly at 8:30 am. If it happens too early, or even just a few minutes late, you risk having a considerably upset listener. Predictability is crucial in making our audience feel secure and in control of their environment and information intake.

News is also a valuable sponsorship opportunity on air, through social media, our website, and in business, weather, and traffic reports. However, I often wonder about the predictability of weather and traffic reports.

They seldom have any genuine interest unless you have a traffic news reporter on the scene, who phones in live and warns us, *"We have a million-dollar Ferrari burning on the corner of Fifth and Main. That will result in considerable delays. But it is worth the wait to see that much money going up in smoke.*" That is an informative, captivating, and somewhat amusing report.

As a "breaking news" source, breaking into a regular non-news program and briefly and concisely reporting a significant breaking event is acceptable. This also enables you to forward-sell the following bulletin, which can provide further details on the breaking news announcement.

Also, remember that all news, weather, business, and other features are always "XYZ-FM News, sponsored by Coca-Cola." It is OUR feature, not Coca-Cola's.

My favorite talk and news guru is Valerie Geller. I highly recommend her books "Powerful Radio" and "Beyond Powerful Radio." (https://www.beyondpowerfulradio.com/).

I read them probably twice a year, every year.

Forward-Selling

Forward-selling opportunities or "making appointments with your listener" should be seized at every opportunity. Forward-selling significantly increases the likelihood of extending listening time and increasing cumulative audience numbers.

Forward selling *"Coming up in ten minutes is our eight-thirty news bulletin. The bulletin includes an announcement from the President that will directly influence your child's future,"* will almost certainly ensure that all parents remain tuned in to hear that bulletin. Even if you are not a parent, you will hang on to listen to the announcement.

Forward selling, such as saying, *"You are going to hear these three great songs in the next fifteen minutes,"* followed by playing the hooks of those three songs immediately preceding a spot break, will significantly assist in holding your audience through that spot break.

The perfect tool to achieve this is to drag the upcoming songs from the log into RCS's Zetta Stacks, then segue and play the three hooks as if they were a produced element.

Forward selling includes selling the next fifteen minutes of your show, the remaining highlights of your show, what extraordinary will happen tomorrow during your show, making an appointment for the next show, or even making an appointment for another show in the future.

Including at least two to three forward sells every hour is a common and sensible practice. These should be scheduled into the log and not left to the on-air presenter to remember. All of these contribute to increased time spent listening.

Listener Telephone Interaction

The subject of telephone calls could be the basis for an entire thesis of its own.

Many of us invite listeners to participate in phone-in programs. However, almost without exception, most callers are less than informed or amusing.

To avoid boring conversations, record all the calls off-air. Cherry-pick the best parts of each call, edit them into a single event, and then play that out. RCS's Zetta's Quick-Record and Stacks are the perfect tools for this.

In radio broadcasting, regardless of format, delay systems (often referred to as profanity delay units or dump boxes) are crucial tools for live call-in shows, enabling a host or producer to intercept and remove offensive content before it airs. These systems typically offer a 5–10 second buffer to give operators time to "dump" inappropriate material.

Large commercial or public broadcasters almost always go with the Eventide BD600+. Mid-size or budget-conscious stations often choose the Henry Engineering Digital Delay II, and digital-native or AoIP setups may use integrated Telos/Axia solutions.

If your station does not have such a device, get one. And be sure that everyone working on-air knows how to use it with muscle memory. A 10-second response buffer goes past you quickly.

In the case of a News/Talk format, have your producer thoroughly screen the caller for content and quality before connecting them to your studio. Remember that the presenter always controls the caller: the caller never controls the program. As a PD you must pay special attention to that.

All your call-in listeners/contributors must be interesting or amusing, or they don't get on the air.

Rule: Only broadcast compelling programming when it involves listener interaction on-air.

Podcasts

Note this legal statement. I am concerned that in many countries, particularly in the USA and the European Union, specific copyright regulations apply to podcasts and music streaming. You need to investigate these before you take the plunge:

> I am not a lawyer and do not have any formal training in copyright law. I present this information for educational purposes only. The information I provide is an opinion and not professional legal advice. Concerning copyright material you wish to use in a Podcast, please seek an experienced copyright lawyer's advice.
>
> Particularly, take legal advice on copyright when using third-party information or commercial music that you do not personally own. I am not liable for any damages resulting from the use of this media or information.
>
> At a minimum, full attribution to the source is suggested.

The mechanics of building and distributing a podcast are widely available. We don't need to discuss those.

Like radio, podcasts pose a common primary challenge: understanding your audience and their preferences.

Creating a podcast solely based on your interests may not ensure its success.

Therefore, conducting thorough audience research to determine what your target audience wants to hear is crucial. But it is much more than that.

It requires singular dedication to focusing intensely on what your audience wants to hear, rather than the story you want to tell.

No amount of advertising spend has the same value as a personal recommendation from your fans. Successful podcasts resonate with a core audience, engaging them to the point where they become avid disciples of the podcast. Word-of-mouth promotion is the most potent tool for podcast success.

Like radio, your podcast needs to appeal to a community of like-minded people. You need to ask: "Who listens?" "Why do they listen to my podcast?" "What do they want to hear?" And again, continuous ongoing research is the key to success.

Then, you must establish how to make money from that podcast, and that's not so simple unless you're able to amass millions of plays.

Let's step back to when radio featured drama series and plays for a moment. Those were the original podcasts.

We listened week after week because the subject, characters, and presentation of the programs gripped us sufficiently, so we made sure to tune in again next week at the same time.

Good talk radio also has potential for engaging podcasts, as evidenced by James O'Brien's excellent daily podcast on LBC in London.

For many years, I tuned in to a particular presenter on one station whenever she interviewed someone from any industry, art, or craft. Her research about that person and empathy for the interviewee were so compelling that every show virtually compelled me to listen.

So much so that, on one occasion, I phoned a person I had an appointment with to apologize and let them know I was stuck in traffic. Not true. I was parked outside his office, enthralled and compelled to listen to the interview's close. My arriving half an hour late was a testament to the presenter's ability to have me commit to being late for an appointment on purpose.

And, yes, I loved her interviews, and I often spoke to many people about what I had heard and learned, as well as what I agreed with or disagreed with.

Howard Stern's programs, for instance, are a prime example. Despite some content that may be considered crude or over-the-top, Stern and his team's sheer ability to captivate their audience is undeniable.

Also, I often listened because Howard's craftsmanship enthralled me; he is the consummate professional talk show host.

Not all podcasts may align with your values or meet your standards of acceptability; some may even offend you. Challenging your audience's norms can open their eyes to a different, more comprehensive, and perhaps perfectly valid world, even if it is offensive or shocking.

When it comes to music podcasts, the challenges are specifically designed to meet the needs of a particular audience. Anything less, and you may find yourself with an audience consisting solely of your loyal family.

On the "family" issue, listen to "The Rock Professor's Podcast" at https://podomatic.com/podcasts/krips48, or on Apple Podcasts, iHeart, TuneIn, Deezer, and wherever you usually find good podcasts. These rock music podcasts have consistently been among the world's top three rock music format podcasts for many years.

He has a laser focus on a particular and rabidly loyal audience, whom he constantly monitors to ensure he delivers what they expect.

He regularly appears and teases on LinkedIn, Facebook, and across the entire spectrum of social media.

You can create podcasts from your news bulletins and make them available on your website, and these podcasts can be associated with in-depth coverage (sponsored, of course) of the stories covered in the bulletins.

You can also quickly generate podcasts from your programming, add open and closing billboards, and post them as features on your website.

If you are doing this, ensure that you frequently promote and guide your audience to these resources.

Tens of thousands of podcasts are available, but only a few have been specifically tailored to a target audience and met their needs. Even fewer are adequately promoted on air, social media, or (gasp) in the press or on billboards.

Again, Brad March has firm ideas on podcasting:

Radio must now contend with a variety of platforms, including social media channels like TikTok and Instagram, as well as streaming services like Spotify and Apple Music, which offer consumers convenient, often ad-free music options. While podcasts may still be behind in revenue generation, they're also vying for listener attention.

Most podcasts revolve around either specific topics or are driven by personalities, but being a star alone isn't enough; the content must be engaging, relevant, compelling, and fresh. The same rules apply. It's likely we will see more crossover between podcasters and radio, similar to what has been done by the We Mean Well team and the Life Uncut podcast duo, Brittney Hockley and Laura Byrne, both of whom are finding success on radio.

Podcasters like Toni Lodge and Ryan Jon have emerged as global superstars, establishing their brand "Toni and Ryan" by producing content that is relatable, entertaining, and unique, targeting a specific market segment. Importantly, their content is compelling, and they know their audience well. Their success is evident, having amassed over 100 million downloads and is gaining popularity not just in Australia but also in the US, UK, and beyond.

They've skillfully utilized social media to promote their daily podcast while also enhancing their brand and entertaining their audience.

Internet Streaming

The previous legal notice also applies to internet streaming.

It is a given that you must stream your primary programming.

Most countries these days have a vast worldwide diaspora, and these expatriates are prime candidates to listen to "something from home."

And with the advent of Android Auto and Apple's iOS CarPlay, many listeners prefer to stream their favorite radio station via Bluetooth to the car audio system because of the other rich feature set that these apps have, which a regular car radio doesn't—features including GPS navigation, browser search, music playback, SMS, and telephone.

The internet also allows you to stream to highly focused subsets of your listeners at very minimal costs. Using subgroups like XYZ-FM's Golden Oldies, XYZ-FM for Lovers, or XYZ-FM Party Time enables you to target and monetize precise, desirable segments of your typical demographics. Streaming these channels on your website or on apps like TuneIn is now technologically very easy.

Again, research the numerous streaming opportunities available to your station before you commit to any platform.

Expecting that your potential audience might magically find these streams is delusional. Make sure that your potential audience is aware of their existence.

Promote them extensively in as many media outlets as possible, especially on your station, social media platforms, and website. And dare I mention newspaper and magazine advertising, too? Everything has its place and its following.

If you need affirmation of the value of a highly active and promoted website, data for the first three months of 2024 in South Africa revealed that Jacaranda FM in Johannesburg was exposed to 564 million website users. Accessing Jacaranda's internet stream via the website or their app reportedly resulted in over one million monthly Internet streaming sessions. They're impressive numbers by any metric.

AI-Generated Presenters

When first introduced, voice tracking was widely criticized as potentially displacing jobs and rendering presenters redundant. It didn't happen. Properly voice-tracked shows are often far slicker, more engaging, relevant, and more professional than "live."

The ability to use Artificial Intelligence (AI) to create presenters is on our immediate horizon. AI is widely decried as the "beginning of the end." Perhaps that's not the case. Approach it as just another tool in your arsenal.

An extremely professional, compelling, and informative AI presentation can be achieved where appropriate. As with all AI models, the key to success lies in the way you prompt the AI agent. That is a skill that everyone in all walks of life will need to learn very soon, and there are many tutorials available to you, particularly on YouTube.

Using The Tools

Modern programming tools like GSelector allow you to schedule based on mood, energy, tempo (open and close), texture (open and close), artist separation, group, local content percentages, era, genre, sound code, instrument, theme, vocalist, participants, song title, and … much more. The choice of rules and goals can be bewildering and overly complicated.

The joy of a well-balanced database populated with the correct mix of artists, genres, sound codes, energy, mood, type, tempo, local content, and much more is that, regardless of how you sequence them, it will sound good.

All you have to do is decide what is most important in your vision and implement it. Don't overthink it; keep things simple.

Your listener figures will soon tell you if you are doing things wrong. If you are doing things right, you will form an emotional bond with your listeners who hear music and programming that defines them, and on-air talent that they empathize with and welcome into their lives. ‘

Brad March comments: *However, there are certain market nuances, songs, and features that are best handled by a human being curator.*

The great thing about GSelector is that it has reduced the time needed to manually manage music logs from 6-8 hours a day to just 1-2 hours a day. I would suggest that you can achieve music logs that are about 70-80% correct as long as GSelector is set up correctly, without any manual scheduling.

Deeper Into Scheduling

I have had the privilege of training RCS' clients in many countries around the World and working with the most talented programmers in the industry, as well as at numerous stations in Africa, Europe and Australasia. Whenever the station and program management make an opening statement, it is often, "*We do things differently here*", which tells me that they have a problem we need to help them resolve.

Based on listening to, reviewing, and working with some of the most successful stations worldwide, I have observed that successful money-earning stations follow similar practices regardless of country, culture, or language. "Doing things differently" deserves cautious scrutiny.

Yes, scheduling a music, talk, or music TV station is a complex blend of math, art, and instinct, but it is neither trivial nor a part-time job. It takes intense work and can become incredibly complex. Much of it involves tedious attention to detail. Nonetheless, it is not magic, although I often see how GSelector delivers precisely what I wish for as Magic.

All of it is based on a firm vision in your mind of who your listeners are, how your station can best address their needs, and the internal balance and natural rotation of all categories and attributes you use.

Start with simple strategies and gradually build your knowledge and sophistication. These strategies have been developed, refined, and tested by successful radio stations and broadcasters worldwide.

With time and practice, you'll find that, while not easy, music scheduling is undoubtedly not a mysterious Black Art. The truth is, there is quite a lot of original thought here. But…

We are not doing brain surgery. No one will die if we schedule the wrong song at the wrong time. "Standard practice" is a well-known and internationally accepted term. Yes, nuances are very apparent, but the basic philosophies are held faithfully, worldwide.

I have gathered generally accepted standard practices and wisdom into one place. Thank you to those I have chosen to learn from and follow.

A good idea is a good idea; reinventing what works for virtually any country, culture, or language is not necessarily a unique or successful programming strategy.

What makes a successful radio station? It is the ability to attract and hold the listenership of a specific demographic for significant periods.

Then, you have a precious commodity: the loyalty of these specific listeners to your station's brand, which you can sell to relevant marketers.

Radio is a business that allows us to indulge in our passion and have fun. However, it *is* a business.

As previously noted, you have two clients and two products. The primary clients are the listeners, and the product you offer them is your programming content. The second client is the marketer, and the product you provide to them is your listeners.

Once you are comfortable and have a reasonable starting point, you can extend your knowledge and capabilities and explore the complexity of more sophisticated and nuanced programming philosophies and tools.

Let's consider the wisdom of Randy Michaels. His insights can provide valuable guidance in your journey to master the art of music scheduling.

Randy Michaels, the programming genius who grew Clear Channel from 200 to 1,200 stations, had an oft-repeated mantra:

"Play the Hits. Only the Hits. All the time."

Building Clocks

No guaranteed ratio or order of categories defines a specific format. What is correct for *your* AC or Top 40 format is not guaranteed to work for any other adult contemporary or Top 40 format.

Randy Michaels scathingly said: *"Slavishly following your consultant's requirement for a power, then two currents, followed by an oldie/classic, followed by two recurrents, is not programming. It is paint-by-numbers."*

Instead, I promote the concept of BALANCE in all relevant attributes and categories of songs in a database and your clocks.

If that balance is true to your vision, then almost any song will successfully segue into other music. However, the position of every song category, live speech, station imaging, Spot block/Commercial break, and promos should be carefully analyzed and placed.

Clock design is where math meets art and magic. There is no such thing as a "perfect" clock. Different listeners on every station at every hour of the day have different needs.

Nor is a clock made for Station A correct for Station B. It doesn't happen. And, a mythical "perfect" Adult Contemporary, or any other format, clock rotation doesn't exist.

It is your station, and it all depends on your vision regarding the ratio of hots, cores, and oldies you play, in what sequence, and when.

Considering all the other elements around it, choosing which category and position is challenging. Additionally, the position of an Opener is always of strategic value. GSelector has the notion of an extra-special best-of-the-best song, known as an Opener.

An opener song is sometimes required for a unique position in the clock: the first song in the hour, the last song before a spot set, the first song after a spot set, or the last song in the hour. You should dedicate these memorable songs to crucial positions in the hour. Additionally, this will assist in keeping your listener engaged into the next quarter hour.

After carefully researching your desired listeners, we may have decided to aim for a balance of three HOTS, seven CORES, and two MEMORY MAKERS per hour.

That ratio is impossible to maintain every hour, because some hours will have only two songs, while others have fifteen.

However, it makes sense, provided you attempt to maintain that general ratio of category exposure over every rolling hour throughout the course of each 24 hours.

The balance of the various categories within the database must remain true to your vision. You are not a nostalgia station when 75% of your database is current.

Do not promise to be a breaking-new-songs station when 90% of your playlist is from many months or years ago.

You are not a smooth Jazz station if you play predominantly R&B. If your music and general energy are within the relaxed and laid-back range and your playlist exceeds one hundred songs, you are probably not a Top 40 station.

Typically, we build categories that represent our vision. If we are an R&B station, then at least 75% of every song in every category in the database must be R&B.

Clocks must represent a template of everything that you want to happen, every hour of the day. And everything means EVERYTHING.

That includes opening and closing billboards, placeholders for pre-recorded events complete with a target duration, live events also with a target duration, station imaging, promos or news, exact time markers, spots, and music—everything.

This is the only way your clock has a chance to represent what should happen on air accurately, and it is the only way you can begin to craft and time each segment and hour properly.

Review the position and every event segue to ensure your vision is being met.

How do we deal with new, potentially hot songs? Generally, they are unfamiliar.

To ensure they are slick segues from song to song, we should cradle and nurture this new song with familiar and loved power or current songs on either side, one before and one after. Do we allow two new and potentially hot songs in a row? No, that's a bad idea.

Do we play two core songs in a row that are on the decline in research numbers? Again, not a good idea. If they are on the way down, consider moving them to a less highly rotated category or resting them for a period.

Rising researched songs must be separated from falling research song categories, always. That way, you can adequately control the flow based on research figures supplied by your audience.

Research carefully, build your clock mindfully, and ignore your competitors' jeers about playing, dumping, or moving songs. Listen to your audience, not the competitor.

Don’t listen to record companies when managing your playlist, either. In an ideal world, you shouldn’t have to meet with record companies, although I appreciate that this is difficult to achieve. Avoid meeting with them, but accept their songs for review at your music meeting.

Obviously, they have a very different agenda from yours, and often, they are superior salespeople who unduly influence you and impact your vision.

Make the right decisions for your audience, and make sure your music rotations make sense to you.

Repeat this mantra as we enter each position in the clock:

- What is the time of day?
- What is your listener doing right now?

- How can we best enhance that experience with what we currently offer?
- What are your competitors doing right now?

It is always helpful to consider what your competitors are doing now. We know that listeners tend to switch from one station to another, often on a whim. The trick is to avoid instigating a switch or be aware of when you might expect a switch from a competitive station. Be obsessive about what your competitive station is doing at any moment.

There was a station in Australia some time ago (let's call it Station A) which actively monitored the market leader (Station B). As soon as Station B went into an ad break, Station A ran a short promo stating that they played more music than Station B, and "If you don't believe us, have a listen to them now. They're ALWAYS playing ads." Station A's listeners would change stations, hear an ad, and immediately come back to their favorite music station and stay there; in the next survey, Station A wiped out Station B.

Historically, we believed that spot/ad breaks caused considerable channel switching among listeners, however PPM research in the USA suggests that the overall average loss is considerably less than expected.

Years ago, we believed that up to 75% attrition occurred by the end of a 3-minute spot block; however, PPM has revealed that this is much lower than that. Something in the region of only 10-15%, if any.

PPM still needs to reveal how many listeners we lost and how many we gained. We could have lost 75% to some stations and gained that same amount from other stations. Nonetheless, we know we can reasonably expect some audience loss, albeit temporary, if we're doing the rest of our job the right way, before, during, and after a Spot block/Ad break.

Naming Categories

You decide how to name each category. If WIZZBANG means a new, potentially HOT/HIT song, go ahead; it's your station. However, some conventions are followed by many formats, and they may also suit you.

Billboard Magazine (https://www.billboard.com) has published a variety of well-respected and carefully researched charts every week since 1936. A subscription to Billboard is something that you ought to have in your arsenal.

By 1950, it had firmly established itself as the primary source of hit parade information for the USA and Canada, and thus for much of the World.

Billboard started defining the chart of song popularity against particular and defined criteria. Billboard now includes radio spins, podcasts, streaming services (both video and audio), and Luminate sales tracking data.

Luminate was developed to track sales of music and video products throughout the United States and Canada. Any music product with a UPC or EAN barcode and ISRC code is eligible for tracking by Luminate.

Luminate's website states, *"Sales data from point-of-sale cash registers is collected weekly from over 14,000 retail, mass merchant, and non-traditional (online stores, venues, etc.) outlets. Weekly data is compiled and made available every Wednesday. Luminate (formerly SoundScan) is the sales source for the Billboard music charts."*

The primary weight of data is radio airplay, streaming spins, and actual sales. From this data evolved the terms "Currents" and "Recurrents." Proprietary formulae remain regarding how Billboard charts and drops songs.

The calculations are presently based on some specific rules. Billboard's rules change occasionally, and it sometimes breaks its own rules. But it is what it is and remains one of the few internationally accepted currencies for measuring a song and an artist's popularity.

These categories are generated for virtually any format.

A CURRENT song is on the relevant Billboard Top 100 chart. A song generally only enters the Top 100 if it gains a calculated position on the chart above number fifty.

A RECURRENT is a song that automatically drops from the Top 100 to become a recurrent after being charted for over twenty weeks below the 50th position.

This rule was further complicated in 2015, as any song that drops below the fiftieth position will also be dropped from the chart and become recurrent if it charts below fifty within the first twenty weeks of entering the chart. And then, in 2015, they introduced another drop rule. If a song charts for any week below number twenty-five after its fifty-second week, it is also dropped from the chart and becomes recurrent.

In theory, and according to Billboard's published rules, a song that has become recurrent cannot re-enter the Top 100 charts.

Well, that's not entirely true. When Whitney Houston passed away, sales surged, and several of her songs re-entered the Top 100.

Confused? Join the rest of us.

However, precise mathematics is involved, primarily based on Luminate, which revealed that a song can remain popular for a considerable period after being dropped from the Top 100. Hence, RECURRENTS.

Billboard practices also define the POWER category. POWERS are generally the top five songs in the current charts. Some stations take it a step further to ensure that all strong-featured songs are played frequently. Up to 30 powerful currents could populate a power category for some stations.

Then, we have the GOLD/OLDIES category. These are older hits that remain of exceptional value to our listeners.

Often referred to as Memory Makers, these songs are a select few that live on for a considerable time, if not forever. We want to hear them occasionally; they bring back memories and joy. However, hearing them too often can burn them very quickly.

Beware of categorizing songs as current, recurrent, classic, or oldies by era. For example, in an adult contemporary format, many songs in these categories might be thirty or more years old.

But, in a CHR/Top 40 format, a song only thirty days from its initial release might be considered an oldie for a fifteen-year-old listener.

Add sub-categories to these, and you can arrive at a bewildering number of categories. Soon, you will run out of space in your clock to fit them in, and the brain's capacity to manage them will be exhausted.

For example, we might decide to classify our categories as:

HOT – very high rotations, repeating many times per day. A top 40 station could rotate a specific song from eight to twelve or more times daily. These are generally new songs, but we have played them sufficiently as rising new or hot songs to make them hits.

RISING HOT/NEW—These songs are in high rotation because they are new songs from familiar and loved artists or exceptional new artists we wish to introduce and make "familiar."

For a station that features breaking new songs, we must consider that all the various new, hot, and rising hot categories are core to our sound and our audience's preference for music. You might repeat every song four to twelve times a day.

Now, we have our CORE categories; these songs define who we are as a station. Also, these represent our listeners; they are the "soundtrack of their lives."

Core categories should comprise at least 60% of your spins unless you are heavily oriented to breaking new music. In which case, breaking music is your core.

In most cases, for anything other than a Top 40/CHR or Hot AC format, you should expect a core category song to rotate at least once daily or every day and a half.

Amongst the core categories, you might find:

- Powers
- Falling Powers
- Rising Powers
- Currents
- Falling Currents
- Rising Currents
- Recurrents
- Older Currents
- Spice Recurrents

Generally, we look to potential, current, or ex-hits as criteria for our playlist. Above all, only play the best of the best songs.

Your airtime is precious; wasting it on very ordinary songs is not doing you any favors.

"Spice" or "Album deep-cut" songs were and remain great, but might not have topped the charts.

Typically, these were on the Billboard Top 100 but might not have reached the top five or ten. A sub-category of Spice is appropriate to Core or Oldies.

Too many categories and complexities will make it impossible to consistently control your station's character and essence. Keep it as simple as possible.

Setting-up Categories

Categories in music scheduling software define groups of songs you want to control regarding how often each song is repeated within a specific period. “Repeating a song” is perhaps counterintuitive. Why repeat the same song at particular periods?

Repeating a great song gives our listeners a warm and fuzzy feeling of familiarity and inclusion, which can lead to considerable loyalty and engagement with your brand.

If that great song is part of the soundtrack of their lives, defining both who they are and the character of your station, there is all the more reason to retain their loyalty.

First, we need to examine our audience's listening habits. Ask yourself and those you work with how often they would play a specific song on their own music player or mobile phone for personal enjoyment.

Generally, this would be a song from a familiar and loved artist. That will hint at how often you need to repeat the same song on your station.

Categories are all about Time Spent Listening (TSL) by your audience.

Nobody, except possibly you, listens to your station 24 hours a day, seven days a week. Instead, they listen for a limited period, usually at the same time each day.

Time spent listening in the USA was previously believed to be as little as seven to sixteen minutes per day. However, Nielsen's PPM suggests this is up to almost two hours daily of TSL for the USA. TSL is much longer in Africa, ranging from one-half hour to four or six hours per day.

Your research, or whatever listener research is publicly available, will reveal your station's TSL.

This TSL is the primary factor around which you should calculate your required rotations, exposure, and, thus, song categories.

What if listeners complain that you are playing a specific song too often? Perhaps they are saying, *"You are playing the **wrong** song too often?"*

It wouldn't be played *too many times* if it were a great song from a familiar artist or even a great song from a new artist that fits into the soundtrack of their lives.

Let's assume TSL of one hour per day. That is one twenty-fourth of your entire day.

So, if you persuade GSelector to play the same song every day but rotate that specific song through a different hour of the day for each day of the week, you can assume that our one-hour-per-day listener will hear that song at least once every twenty-four days.

If we play a new song that we are trying to create familiarity with twelve times per day, every day, then a listener might hear the same song twice if they listen for two hours or more. Is that bad? No. It meets the requirement of becoming familiar with and learning to love that one song.

This bell curve represents a typical song lifecycle:

Typically, given enough exposure, any new song follows that standard life-cycle on a bell-curve in its play history.

- Unfamiliar
- Familiar
- I like it
- I love it
- Wow, it's a HIT.
- Then, if overexposed, it begins to burn and needs to have its repetitions reduced by resting it or moving it to another lower-rotation category.

Of course, despite our best judgement, some songs jump quickly straight from familiar to "I hate it". Beware of these. Only continual research can make you aware of where a song is in the life cycle.

And always keep this front of mind: a new song burns for you and your on-air personnel far quicker than it becomes familiar to the listener. After three months of play and over 500 spins at one station, a specific new song was identified as unfamiliar to their listeners. Possibly rotating in the wrong dayparts? Or not played frequently enough?

The optimal approach is to finely balance each category's attributes like genre, sound codes, era, energy, male/female/boy band, etc; whatever best defines your market-leading format and unique sound that grows your audience.

Many stations define their categories by era. The 00s, 90s, 80s, etc. I argue that these are not optimal. If you define songs as an 80s-era category, for example, do you believe every song in this category is equally strong and deserves the same rotation frequency? It's simply not possible. If you insist on defining categories by era or genre/sound-code, at least three levels should be included in each era: high rotation, slower rotation, and spice.

"Spice" consists of extraordinarily loved and familiar songs that might not have achieved hit status; they were in the Billboard Top 100 but perhaps never actually reached the top five or ten.

Be sure not to overplay spice songs, as that could burn an iconic song. Set the three levels to play in a ratio that satisfies you and your listeners. 4:2:1?

Some PDs use genre or sound codes as a category. Again, this is not optimal. All genres and sound codes should be distributed proportionately across all categories.

If you claim to be an R&B-based station, the highest percentage of songs in every category should be R&B, as I mentioned earlier – around 75% is a good rule of thumb.

Any other spice genres and sound codes must be spread proportionately throughout all categories.

In Africa, many stations follow the public broadcaster model for Gospel music. Public broadcasters have a mandate to cater to listeners from cradle to grave. And, traditionally, they would devote Sundays to gospel music.

For a Commercial station, the question is: Why only play Gospel on Sundays? If the song is worthy of being on your playlist, why can't it play in an appropriate category from Monday to Sunday? Do you believe your audience changes their entire character on Sundays?

Another hangover from PBS. We sometimes avoid certain songs on a Sunday because they might disturb or offend our audience. Huh?

If it offends them on a Sunday, surely it will also offend them on a Monday?

Before we decide on categories and music to populate them, we need to have a solid idea of who our listeners are or who we want them to be, and therefore the music that will most appeal to or attract them.

This, in more detail, defines the music your station will feature. Once you have established who your listener is, you have to decide how to categorize this music.

For simplicity and as a basic example, we will consider a current hit station but with a slant toward older listeners who need nostalgia.

Yours is not necessarily a station that breaks new songs or artists. Instead, you might cleverly depend on competitors to take on that risky assignment.

You should also carefully watch the Billboard Top 40, or your local national chart in territories where Billboard isn't so relevant. We refer to Billboard and YouTube to analyze what we should be playing.

You will conduct, at a minimum, monthly research on the song's life-cycle status.

Category A: HITS: This high-rotation category will feature the most popular current hits, those in the top 10 to 20 of the Billboard or national charts. These are songs that your listeners/viewers want to hear/view often, possibly even as often as once every two to three hours during a daypart. Thus, each song should be repeated at least eight times daily (every three hours).

Category B: HITS lower rotation: This is another highish rotation category that will feature new breaking songs that suit your format or might be hits once they become familiar to your listeners.

You'll probably repeat each song every six hours (four times daily), but always in a different hour of the dayparts. Remember that a song will only become familiar once it has been heard sufficiently.

Category C: CURRENTS: This is your core category, with familiar songs by familiar artists that define your character and sound.

These songs could have been hits now or in the past, and they are highly familiar, appealing to your listeners/viewers.

You'll repeat each of these songs at least once per day, always moving hours and dayparts each time.

Category D. DEEP SPICE/CORE: Spice is also a core category but will feature deeper cuts from familiar artists.

These would be songs that possibly were not Top 10 hits in their time, but are very familiar to your listeners/viewers. You likely repeat each of these songs at least once every two days.

Category O is our OLDIES category: Here, you will feature the iconic songs of your format, those that have stood the test of time and still bring a smile to lips and a glow to your listeners' souls when they hear or view just the first few opening lines. You'll want to repeat each of these songs only once every five days at a maximum.

Finally, while a song rests out of rotation, deactivate it. It will retain all its characteristics and assigned category and can be quickly returned to active status as required. You do not generally delete a song from your database but rather deactivate it unless it is never going to be played again, in which case a deletion is appropriate.

Using these criteria, you can arrive at how many songs are needed in each category.

Rule: Once you have established the required number of songs in each category, that number must not change unless you revise your vision.

This number is based on your requirements of songs per category per hour and required rotation.

Any change to the number of songs in a category will compromise your rotational requirement.

Therefore, if you add a new song to a category, one existing song must be sacrificed and moved to another category.

Rule: Songs are categorized by how often they need to play. Not by any genre or local content or era or time-of-day that it should play.

TIP:

The number of songs in a smaller category needs to be odd.

This will help avoid the repetition of songs at specific hours each day. For small categories, read HELP about KICK.

TIP:

When categorizing a song, at no stage have we yet considered the genre of a song, whether it is an R&B or rock song, nor have we considered whether it is local content. Nonetheless, the relevant ratios of each must be maintained.

Other than "does this song have the potential to be, or already is, amongst the best of the best?"; after that, the only criterion for defining which song category a song belongs in is:

HOW OFTEN DOES IT NEED TO PLAY?

Sticking to our basic three categories of Hot, Core and Oldies, the rotation and number of songs required in your clock's song content might look like this:

Category Name	Songs from this category per hour	Same song repeats per day	Number of calls for this category per day	Number of songs in this category
Hot/New	3	8	72	9
Core/ Familiar	6	1	144	144
Oldies /Memory Makers	3	Once every five days	72	360
Totals	**12**		**288**	**513**

Internationally, the Core/Familiar category could just as easily be known as Powers, Currents and even Recurrents.

Thus, their combined repetitions could range from 3 to 8 or more times per day.

TIP:

It may be valid to restrict some songs for Daylight only plays, and others for Evening and overnight plays. Try not to do it too often, though – if you have that much doubt about a song comfortably fitting into your format, perhaps it's the wrong song.

TIP:

Play FAMILIAR songs by FAMILIAR artists.

One broadcast programmer used a simple mantra for programming:

Play the HITS, only the HITS, ALL the time.

If you play unfamiliar songs by unfamiliar artists, ensure you have a good explanation and defense for that decision.

Station Formats

A partial list of potential formats highlights the importance of researching potential listeners in depth.

This is not an all-encompassing list; dozens, if not hundreds, of possible formats exist, each targeting a specific demographic. The critical issue again is that your research defines your format and vision. Beware of a format that is demographically challenged. Meaning: do not attempt to satisfy hugely divergent demographics and ages. It's not possible.

Ensure it focuses on the listeners you believe are in need, resulting in a product that you can sell to marketers.

Recent 2025/26 Spotify and Billboard research suggests a potential change to radio formats. We are still going to see smaller playlists repeating appropriately, But, my guess is we are going to see far less rigid genre structure across the various more popular formats. There are now fewer groups and more solo artists. Significantly these artists are collaborating with other artist in totally different formats; broadening their appeal and popularity.

Here are the top radio formats in the United States, sourced from the USA Radio Advertising Bureau.

There are a vast number of potential formats, each targeting specific demographics.

- **Country, Classic Country, and New Country** are the top formats in terms of the stations that play them. The format spans from Classic Country to Mainstream, encompassing performers who are considered "crossover" artists.
- **News/Talk**: This is a broad category with the biggest audience in the USA. And one of the most challenging formats to excel at.
- **Sports** are growing so fast that they can be classified as a separate format.
- **Oldies or Classic Hits**: These stations primarily play music which appeals to under-55s. They prefer to be known as "classic hits" stations because "oldies" disparagingly suggests a certain age of listeners.
- **Adult Contemporary**: Often skewed towards women, this format began in the early 70s and has successfully replaced what was once known as Easy Listening or MOR (Middle of the Road) stations.

Many Adult Contemporary (AC) flavors exist, particularly Hot AC which is best described as Top 40 without rap music.

- **Hispanic** or other language/culture-oriented formats: Next to sports, the fastest-growing format in the USA is Hispanic, which includes music, news, or talk formats broadcast in Spanish.
- **Adult Standards**: In the United States, this format is in decline. Artists such as Frank Sinatra, Barbra Streisand, Tom Jones, Neil Diamond, and Kenny Rogers appeal to older listeners; however, advertisers targeting this demographic are increasingly less interested in reaching them. Unfortunately, stations running this format may be following their traditional and aging listeners into the grave.
- **JACK** attempted to be everything to everyone. It played an eclectic mix of songs from the fifties to present, across all genres. Not many JACK statons have survived.
- **Top 40**: Also known as CHR, or Contemporary Hit Radio. Top 40 refers to the Billboard magazine charts, considered the industry's bible.

Artists from CHR's initial exposure tend to bleed into other formats. Typically comprised of listeners between 18 and 34, Top 40/CHR also reaches down into the 10-17 demographic, further complicating the issue. Contemporary Hit Radio, or CHR, has been an essential player in the music industry for many years.

CHR stations focus on playing the latest popular songs that attract a wide audience, typically targeted between the ages of 18 and 34. For decades, these stations have shaped the musical panorama by promoting new successes and introducing listeners to emerging artists.

This influence on musical achievements is not just about playing songs; It also includes helping songs to rise the charts and gain popularity. As a result, CHR has often been seen as a vital source to discover the next big thing in music.

In recent years, the music scene has undergone considerable changes, as reflected in Billboard charts. Billboard tracks the most popular songs in the United States, using a combination of sales, radio play, and online streaming data.

Traditionally, new successes frequently emerge, showcasing a new talent. However, there has been a notable trend towards fewer new successes appearing on Billboard charts. Now, many songs are staying for longer in Billboard Currents. And, CHR programmers have to depend on much older songs for their playlist.

This change raises questions about what is happening in the music industry and how traditional platforms such as CHR are evolving in response.

As streaming services grow in popularity and more people turn to platforms like TikTok, Spotify, and Apple Music, listeners have greater control over their music options. They often prefer familiar songs and artists instead of exploring new ones. This behaviour is reflected in the Billboard charts, where fewer new hit songs are appearing and thus, specific artists and songs remain on the charts for longer.

Also, Spotify research heavily emphasizes how older (Recurrent and Oldies) songs have a much longer life than previously. Older songs resurface frequently due to their high streaming numbers and the nostalgia they evoke. Consequently, CHR stations may feel pressured to play these familiar and older successes to satisfy their audience, rather than taking risks with new music. This means a change in how music and the role that CHR radio plays in determining what becomes a success.

Before CHR, the radio stations often played a mix of newer songs and fewer old releases. As pop music evolved, particularly with the rise of genres like hip-hop and dance music, CHR radio began to reflect the sound of contemporary culture. It became a platform in which listeners could discover new successes and artists, making it a cornerstone of the music industry. The role of CHR Radio in promoting new artists and trends cannot be overrated.

Many musicians have entrusted these stations to reach potential fans. When a new song was recognised on CHR, it often meant immediate exposure to a large audience. This helped new artists climb the rankings and obtain recognition. For example, in the late 90s and early 2000s, many pop stars such as Britney Spears and NSYNC gained their popularity mainly through constant high rotation on the CHR stations. CHR played a crucial role in determining which songs became Hits.

During the peak of CHR's popularity, Billboard's charts had significant importance. The Billboard Hot 100, which ranks the most popular songs in the United States, England and Europe, has become closely tied to the CHR format. If a song entered Billboard's Hot 100, it usually meant strong support from CHR stations. This contributed to stimulating sales and streaming of songs, giving credibility to artists and a strong fan base.

The rankings not only tracked down what was popular, but also influenced what was playlisted on radio.

However, these charts also created some pressure. The radio stations had to maintain their playlists fresh, but they also faced challenges in balancing fewer new potential hit music (Currents) with older successes (Recurrents) that remained popular. While CHR concentrated more on established songs, the new artists found it more challenging to break through. This led to a cycle in which the oldest songs received more airplay than new versions. This trend has been reflected in the Billboard rankings over time, showing a decrease in diversity among new successes and a growing reliance on songs and artists that listeners already love.

With the changing times, the effects of this dynamic have begun to shift, leading to a decline in the impact CHR has on new musical successes; its influence is fading.

Several factors have contributed to this decline, significantly impacting the way new hits appear on the Billboard charts.

The rise in digital streaming and alternative platforms changed the musical landscape. Services like TikTok, Spotify, Apple Music, and YouTube offer listeners more choices than ever. People can easily find and listen to music they like without relying on what CHR plays. Instead of connecting to the radio, many listeners now prefer to create their own playlists or discover songs thanks to personalized recommendations. This change has made it more challenging for CHR stations to introduce new music effectively.

When people can access large libraries of songs on demand, they often opt for new over familiar songs or discover their own hits on various streaming services. Many of these new artists and songs are never heard on CHR, which reduces the potential that new songs and artists become successes.

Today's audience has a shorter attention span and tends to listen to songs they already know and love. This is especially proven by Spotify research where even the sixties and seventies groups and artists are appearing more often. Many listeners gravitate towards nostalgic songs, which means that the older successes often dominate their playlists. Consequently, the new songs have more difficulty entering Billboard's Top 100.

With the decreasing role of CHR Radio in promoting fresh music, fewer new singles reach the public. The charts are congested with older songs that people recognize, making it more difficult for new artists to obtain the same level of exposure.

Instead of spotting a rising star or a new genre, the listeners hear the same successes several times. When Billboard compiles their charts, this reflects this behaviour. They capture the reality that streaming numbers are more and more dominated by older tracks, instead of presenting new music that deserves attention.

Or, new artists and songs are only heard on streams and never gain airplay on CHR. This is typically shown in the fact that an artist like Bad Bunny, with 19,8 billion streams in 2025 on Spotify only occasionally appears in the Billboard's Top 100, CHR's basic format.

CHR Playlist strategies have also evolved in response to these trends. Many stations opt for a "safe" playlist filled with songs that have already proven successful. In doing so, they aim to keep their audience engaged; however, this strategy often results in neglecting emerging artists and new sounds.

This lack of variety on air limits exposure to more recent songs, which makes it a cycle that feeds continuously. This is so evident that Current Hit Radio is often referred to as Re-Current Hit Radio.

Given all these factors, the decline in the impact of CHR radio on music hits is evident.

With the rise of digital streaming, changes in the listener's behaviour, and conservative playlist strategies, fewer new successes are heading for Billboard's charts. This drop marks a significant shift in the music industry and the way new music is promoted. Even as the industry evolves, the consequences of these changes will continue to reshape the relationship between artists, listeners, streaming, radio, and Billboard. In recent years, Billboard has shown a clear trend of fewer new hits reaching the charts, and artists holding significant share of the Billboard Top 100 for much longer periods.

This change can be understood by analysing the statistics that detail the number of songs that reach the Billboard Hot 100. A few decades ago, it was common to see many new songs going on the charts each week. New artists and fresh sounds were always being discovered and celebrated. However, the trend has changed over time. With each passing year, the number of new hits has decreased.

Instead, we now see the same songs and artists dominating the charts for longer periods. This lack of fresh music and artists raises questions about where the next generation of hits and artists comes from.

The Top Ten this week is dominated by Taylor Swift (two songs) and Bad Bunny (three songs). Billboard has recognized this trend as a dynamic factor in its classification system, with older songs often appearing on Top 100, sometimes surpassing the latest tracks. And, CHR has seen its role significantly decrease.

This shift indicates a removal from the exploratory nature of the musical discovery that CHR once supported.

Artists may need to develop new strategies to promote their work, as traditional channels such as radio become less effective.

Musicians must embrace alternative ways to interact with their audience, which could mean using platforms such as Tiktok or Instagram to show their work instead of relying exclusively on radio. A typical case study is to be found at mathsartormagic.net on the blog: ***How does Radio discover new music?***

The decline of the impact of the CHR shows how the music industry is evolving. With the tendency towards fewer new successes and a greater focus on older songs, artists and music lovers must navigate a different arena.

This new landscape may seem demanding, but it also offers a possibility of innovation and creativity, allowing a wide range of voices to emerge in the world of music.

Can CHR buck its dying trend? Curated radio remains very dominant if it has strong and relevant presenters with compelling stories to tell. However, streaming is here to stay, and it needs to be carefully countered.

The belief that young people have abandoned listening to the radio to download music is an unproven myth. Nielsens's PPM proves that people may not listen as often, but they still do. Currently, radio is where people first hear the songs they're playing on their music players or mobile phones.

- **Soft Adult Contemporary**: The home of Celine Dion and Adele, this format is among the most popular subgenres of adult contemporary, appealing primarily to women.

- **Religion**: This format can include anything from popular Christian music to teachings, Catholic masses, gospel, firebrand-type ministry, Islamic-centered, or other specific religious persuasions.

- **Classic Rock**: This format is reportedly in danger of dying as traditional baby boomer listeners age, but personally, I doubt that. The music remains exceptional, even some songs that are fifty years old. New, younger listeners are discovering these songs all the time, often by listening to their parents' old record collections.

  Artists such as Led Zeppelin, Pink Floyd, the Doors, the Beatles, Phil Collins, and U2 are the heart of classic rock.

  Many Classic Rock stations' playlists overlap with the Classic Hits format, giving it the same demographic.

- **Adult Alternative,** often known as Triple A (AAA) or Adult Album Alternative, is a radio format and music genre focusing on a diverse, mature blend of rock, indie, folk, singer-songwriter, and roots music. The format typifies my contention that we are seeing a distinct change in format philosophy.

We are still dealing with tight rotations of less than 300-500 songs, but AAA is no longer format bound.

The format features an eclectic, non-mainstream playlist, emphasizing album tracks over singles and appealing to an adult audience. It includes genres like country, indie rock, Americana, blues, folk, and modern rock. AAA features a huge variety of formats, but unlike Jack, AAA retains cohesion and sound.

The focus favours artists with introspective lyrics and mature, polished sounds (e.g., Mumford & Sons, Dave Matthews Band, Lorde). AAA typically avoids heavy metal, hard rock, and rap. AAA often mixes new, emerging artists with deep tracks from established, classic artists. The format features artists like Eric Clapton, John Mellencamp, Tom Petty, The Rolling Stones, and Bruce Springsteen, along with contemporary indie artists.

AAA often breaks new artists and bridges the gap between adult contemporary and alternative rock, focusing on musical craft. AAA perfectly demonstrates how specific artists hold their place in the top hundred, often for 52+weeks.

Local Content

The station manager, program director, or scheduler's primary job is to ensure that you do not lose the right to broadcast by not meeting the terms of your license. Play the required license-mandated percentage for local content, because your station *is* monitored from time to time, and penalties can be significant.

Record companies will argue this, but radio, even more so than any digital medium, creates hits and introduces talent to listeners, thus expanding our cultural comfort zone and fostering the growth of artists and the industry.

We must play and promote local artists and content; however, having a local content category again suggests that all local content has equal value and weight. That's never true.

The optimal solution to local content is to place the relevant percentage of songs mandated in your license into every category.

Then, set local content as an uber-important priority, and GSelector will ensure that the required percentage is scheduled.

There is a cast-in-concrete concept for categories. All songs in a category should receive equal plays compared to their peers over a specific period.

If every song in the category is subjected to brutal scrutiny and is believed to be the best-of-the-best, why would they not be entitled to the same number of spins as their peers?

Be aware that excluding a song from a specific daypart may have unintended consequences. It is immediately compromised in terms of equal plays to its peers. Also, when you exclude a song, you are affecting fair plays to the artist, too, and to any of that song's attributes.

Changing just one song in a high-rotation category can significantly alter the flow and character of your station.

Who is the Star?

"Who is the STAR? The music or the presenter?" Your vision needs to be very clear on this issue, as your presenters may attempt to build their brand rather than your station's.

We open with this comment from Brad March: *Any radio station these days relying solely on music as the main draw is likely to face challenges. Certain formats, like Soft AC (for example, Smooth) and Gold, have such a strong and unique music format that they can thrive without needing Breakfast and Drive entertainment shows. However, in environments where there are multiple competing formats, having talented Breakfast hosts can make a significant impact*

Radio requires an excellent presenter who can maintain energy, has been sufficiently prepared to deal with a wide range of subjects, can listen and mediate rather than pontificate on their own opinions, can bring order out of chaos, and can add amusement and information to the process.

Howard Stern and his team, for example, reputedly rehearse at least three hours before each show and then have what he refers to as an "autopsy" for an hour after each show.

Spontaneous, yes, but within strict pre-rehearsed parameters. Howard Stern, above all else, is an absolute master at forward-selling the next portion of his current show or tomorrow's show. Continually, he teases and hints at what is coming up. The result? Phenomenal time-spent-listening numbers.

Often, the best talk show hosts come from different non-radio careers. They are found, nurtured, and mentored to become presenters.

They are usually brilliant, well-read, highly opinionated Type-A personalities with an off-the-wall sense of humor and powerful personalities. So, they are challenging to control.

As a Program Manager for both talk and music radio hosts: *"When you grab a tiger by the tail, you need to have a plan for the teeth."*

Curated Music Radio, which is often under attack by various media and information sources, relies heavily on the presenter's contributions. Having a live presenter offering relevant, engaging, compelling, and amusing information and commentary is one of the primary differentiators between radio and internet streams.

However, your morning drive show presenter telling us how intoxicated he became the night before and how hungover he feels this morning offers little value to the station and its brand. If anything, it could damage your brand. Yes, I heard such a conversation on a major station's drive show the morning I wrote this.

The golden rule is that a presenter/DJ *never* talks exclusively about his/her own opinions or experiences and is *never* bigger than the station. Except in USA, far-left and far-right talk shows, where all bets are off.

The reality is that despite a presenter's confidence in their personality and belief in the infallibility of what they have to say, not many presenters can be engaging, amusing, informative, and compelling for more than a few minutes, at best.

Of course, there are exceptions. I heard one presenter, Kevin Savage, an absolute master of the art, talk for almost thirty minutes at an outside broadcast. He was compelling, sliding effortlessly from subject to subject, constantly engaging, and often humorous. Why did he talk so long? Because the engineer had asked him to keep going while they sorted out a technical fault at the site.

Never underestimate the value of curated music accompanied by an intelligent and amusing presenter. This is one of radio's significant differentiators from our prime competitor, internet streaming music services.

A few stories about presenter behavior. First of all, in the late 90s while driving across the heart of America after midnight, I was listening to an adult contemporary station. The presenter was both laid-back and professional.

He introduced a song with: *"I was walking down a very lonely street late at night. A tall, elegant, and confident young woman dressed in a long, flowing red dress approached me.*

We approached, stopped, and looked at each other. As a car drove past, blaring this song, I immediately knew I had just met my wife-to-be."

All this is over the intro to Chris De Burgh's Lady in Red. His chat was perfectly timed to hit the first vocal. And then, over the outro, he said: *Chris de Burgh explained how he came to write this song: "I was trying to get the lyrics for a song I was writing.*

And then, across the room at a party, I saw my wife dressed in red. She turned to me and smiled that secret lover's smile. Now I could complete the song."

The presenter owned that song, and he owned me, too. True or not, he revealed something of his inner self, drawing me into his private circle. He also gave me some interesting information about the song lyrics.

Second, at a privately-owned station in a small town (250,000 population) north of New York.

I visited to observe a real-world presenter at work, ensuring we were doing things correctly regarding our Master Control software. Jim was the station's owner, a mature gentleman with a spring in his step and a sparkle of humor in his eyes. He had hosted the morning drive show for twenty years and had agreed to let me sit in.

During the show, which I could only define as "controlled chaos," there was an avalanche of differing opinions, phone calls, messages stuck on the studio window, and faxes. Subjects flew in and were debated, some quite hotly, and then he smoothly moved on to another topic. Only two songs were played during each hour.

I asked Jim how, after twenty years behind the mic, he managed to maintain the incredible energy and range of topics that his audience responded to so enthusiastically. *"When we finish, join me for breakfast."*

Jim had enjoyed breakfast at the same diner, just a few blocks from the station, for twenty years. We entered, sat down, and the controlled chaos continued.

Other customers approached us, slapped Jim on the back, and congratulated him; some objected strongly to his stance on a subject and offered an alternative viewpoint; others demanded that he discuss a specific issue that was bothering them.

Jim said, *"There you go. That is how I maintain the energy and find subject matter important to my listeners."*

Jim was a perfect example of an, unfortunately, dying breed of radio personality: informed, involved, amusing, relevant, captivating, and giving a voice to a community of like-minded people.

Thirdly, a story from one of the top stations in New York. I was visiting the program manager.

While chatting, the station's off-air receiver burbled in the background. Suddenly, he stood up, excused himself, and disappeared for five minutes.

When he returned, a different presenter was on air. The explanation? *"I told him I would fire him if he didn't keep to a maximum of one and a half minutes' chat between songs."*

And finally, this one attributed to Randy Michaels. True or not, I cannot comment, but I have heard it from several sources.

Randy reputedly suffered from insomnia. Being the committed, educated, and passionate radio person that he was, when he was awake in the middle of the night, he would open his phone book and, at random, call one of the Clear Channel stations' on-air live studios around the country on the studio hotline.

He would open with, *"Hi, this is Randy Michaels. What went wrong with that last break?"* An outrageous start to the conversation? Probably not. And it certainly kept the DJ on his/her toes. Every break deserves a review, and every break could perhaps have been improved. This segued into a Master Class on presentation by the guru Randy Michaels.

The Intimate Details of Programming

All of these philosophies are implemented in RCS's GSelector. It's well-known that men are reluctant to read manuals, but I suggest that, like sensible women, you should read the in-line context-sensitive Help in GSelector.

GSelector's Help manual is an incredibly detailed work of art. Do use it. Additionally, I strongly recommend that if you are new to GSelector, you request training from one of your local RCS Support Engineers. Once you feel you have mastered GSelector, it is time to request training again. You will be amazed at the capabilities hiding "under the hood."

The RCS Support Engineers are generally from a broadcasting background; many are ex-program managers with years of real-world experience. They have a deep understanding of programming and the nuances that GSelector can empower you with.

We always start the process of setting up and training a new station with a discussion on the five Pillars of Programming:

- Transmitter footprint
- Who is our listener?
- What will we play or do to entertain, inform, and amuse them?
- How to build Time Spent Listening (TSL)
- Research

We then discuss our vision for the station in detail. Hopefully, this will lead to a written program guide that everyone can work with.

We set up all the variables and the look and feel of GSelector.

We discuss the music, categories, and expected rotations and repetitions of each category's songs. Repetition is all about how often we want a specific listener to hear each song in each category.

We establish the categories we will start with. Then, populate each category (GSelector and GLinker) with at least one event to establish clock timing.

We cannot yet decide how many songs will be required in each category to achieve our required song turnover durations.

Next we build and assign clocks twenty-four hours a day, seven days a week. This process also promotes discussion about exactly how each hour will flow, including what imaging will be used, where the spots will occur, what features may need to be scheduled, and why we are playing a specific category of songs at each position, among other topics.

We don't yet consider rules and goals, nor do we deal with clock goals or constraints; those come later.

We pay attention to the required Hots, Cores, and Oldies ratio. Of course, you are not limited to these specific category names, and we are not slaves to those ratios. The primary criterion is *"What do we need to hear now?"*

Once the clocks are done, we can start populating the various categories. As we do so, we aim to achieve a balance in every category of all the criteria that are important to us.

This includes the station's *sound*: sound codes, energy, era, local content if required, and other artist and song attributes.

As we add all these elements, GSelector calculates the predicted natural song turnover for each category based on the clocks we have built and the number of songs in that category.

We continue to add songs to a category until we reach the optimal number that, against the clock, will result in achieving our desired song turnover and repetition in line with our vision.

Now, to change the natural rotation of a specific category, we can either adjust the number of songs in that category or the number of calls in the clocks for that category.

We set up the starter pack of goals and rules and start scheduling. We can now review the results of our efforts and identify any issues that require attention.

And then, only then, we start to nuance the way the programming is occurring, fine-tuning to as close a representation of our vision as possible.

Time Spent Listening/Viewing

Remember that the average listener only listens for part of your 24-hour broadcast. They are more likely to hear during a specific daypart each day, such as while getting ready for work, driving to work and then back to home, or late in the evening while winding down.

It is doubtful that they listen for extended periods, although running something like the "At Work Network" or "Office Blocks and Home Units" from 10 am to 3 pm to encourage listening at the workplace or while working at home is a proven formula that builds daytime numbers. Both features involve a lot of audience participation, song requests, and contests, and this can be applied particularly well to the variations of AC formats.

In South Africa, the average Time Spent Listening (TSL) is approximately one to three hours, although some stations claim up to six to eight hours of TSL.

That extended TSL would require a very different natural song rotation and number of songs in each category than a station with only one-hour TSL.

We want to ensure that we don't deprive any of our listeners during the time they spend listening to us, regardless of the hour or day they tune in. They all deserve the opportunity to hear all our best-of-the-best songs.

The number of times a song is played and repeated per period should be considered with TSL at the top of mind.

We are attempting to rotate each song through every hour, day, and part of the day. For example, let us look at a listener who we know listens from midnight to 3:00 AM every day, whilst on shift at the local fire department.

This three-hour period represents only one-eighth of our day. So, if a song is playing only once per day (every day), our hypothetical Listener/Viewer may only hear/view this song every eighth day. Thus, a song playing once per day is rotating (as far as this listener is concerned) on an eight-day cycle.

Similarly, a song repeated on our station every five days may only be heard by this listener every fortieth day.

Remember TSL when you consider song rotations. What might initially sound like an extremely high (frequent) rotation might be low for a specific listener.

Goals, Rules and Priorities

GSelector is unlike all scheduling competitors. In all other scheduling products, you set up rules to establish what you **DON'T** want to happen. You don't want the same Artist to appear within xx hours of itself. You don't want more than two R&B songs per hour, and they must be at least three songs apart. You don't want a high-energy song followed by a low-energy song.

In GSelector, all the much-loved and used legacy rules of Selector are available. But GSelector, the "G" denoting GOAL Selector, is quite different. GSelector enables you to establish goals for what you **DO** want to happen.

A considerable number of rules and goals are available in GSelector and GLinker.

However, if your music is chosen correctly and balanced, you will not need to apply too many goals and rules to ensure a smooth music flow and maintain constant control of your station's character.

Always start a new station with the absolute minimum number of rules and goals implemented, allowing GSelector to run the algorithms unhindered and “breathe”.

We only introduce additional rules or goals if we identify a problem that needs to be addressed.

There are two kinds of rules that you can set.

UNBREAKABLE rules are all equally important and are, by definition, unbreakable. If a song fails to meet any unbreakable rule, GSelector will not schedule it in that position.

BREAKABLE rules are always listed with the most important rule at the top and the least important at the bottom. Breakable rules are applied to the most rested stack from the bottom (least important rule) to the top (most important rule). Ultimately, a song that breaks no unbreakable and the least breakable rules will be scheduled.

Legacy (from the original Selector) rules are only required to address specific requirements or problems.

They should only be implemented after thoroughly reviewing the balance of each category in your library.

If you hear too many reggae songs in any hour, this is almost certainly due to having too many reggae songs in some or all categories.

A basic starter pack of goals and rules is:

LOCAL CONTENT: In countries where Local Content is required this is an uber rule, above and more important than all others. You set your required local content rules to your license-mandated percentages (minimum and maximum). If the balance of all your categories meets this requirement, GSelector will ensure this is met in your programming. The local content rule helps ensure you meet your local content requirements, and a handy tool—Dynamic Statistics—keeps you aware of the exact status when editing your log, as well as your local content percentages, hour by hour.

We have a further option of unbreakable and breakable goals and rules.

I initially suggest no unbreakable rules, other than perhaps a Yesterday rule that avoids a song playing in the same hours, n-days backwards. N-days are interesting, too, in that you can look back multiple days.

This should not be required if you are using correct Search Depth. Read the manual.

You can define how concerned you are about "yesterday" vs. "three days ago." Generally, we are 100% concerned about the same song playing at the same hour yesterday, 80% worried about the day before that, and only 50% concerned about three days ago. Very few people can recall the last time they heard a song after three days.

Breakable would be, in order of priority:

CLOCK GOALS: On the clocks, you can define particular goals or constraints for a specific song position. For example, I want the general energy of the music to remain at a minimum after midnight and before 5.00 am.

But on Friday evening at 10.00 pm, its party time, so I would want all the songs' energy to be at maximum, all dependent on the balance of energies available in the library, ranging from 1, which is categorized as laidback and doped, to 5, which is head-banging and off-the-wall. Or however you describe the energy of your library.

Additionally, you can set constraints based on specific attributes of the songs, such as requiring a song position to have a certain tempo or a particular range of BPM, genre, or characteristics, like male or female vocals.

Then we define clock goals based on constraints at specific positions in our clock.

For example, this song must be energy five, with a sound code of Hip-Hop and Female. Or, if I wanted an R&B binge, I could require that the following five songs must be R&B.

SONG SPREAD: This ensures that the same song repeats as far apart from itself as possible within the balance of the categories.

ARTIST SPREAD: This spreads artists as far apart from themselves, again within the balance of the categories. This caters to the fact that you might have the same artist in multiple categories. Maximum artist spread is maintained, but you must always ensure that the artist's name is spelled correctly and consistently.

SOUND CODE SPREAD: We strive for sufficient exposure to all sound codes in your playlist rather than clumping them together.

Unless you want them together, the higher-priority clock goal constraint will apply and override this rule.

ENERGY SPREAD: I prefer using Energy to Tempo.

It gives more control over the "feel" of an hour, and I don't have to set up all sorts of tempo rules to avoid clashes.

With energy spread, I find a generally delightful hills-and-valley flow to the programming, with neither too high nor too low energy.

DAYPART GOAL: This differs from the other legacy daypart rule, which ensures that if a song plays in daypart A, it will only repeat there once it has played in a defined number of other dayparts.

In GSelector, the daypart goal is much more refined. It ensures that if the song is played in the first hour of a specific daypart, it will only repeat once played in the other hours of that same daypart.

- **Rule** - GSelector will only respect those Goals that are set in the Policy Priority Rule assigned for each category.
- **TIP** - Each category could require a different set of goals from other categories.

- **Rule** - All clock goals (including clock constraints) should be set at the highest priority, unbreakable position.

Projected Turnovers

Most rotational rules and goals are based on a song's natural projected turnover. The natural projected turnover of each song is the Holy Grail that schedulers strive for. This results in every song being played equally to its peers over time.

Projected turnover is calculated by reviewing all clock calls for specific categories over time. In GSelector, this is taken down to the Daypart level.

This projected turnover will tell you how often the same song in a category will repeat, assuming it is always scheduled and never skipped due to failing any of the rules or goals you have set.

Be aware that Search Depth has significant impact on actual turnover. Read the manual!

Specialist Programs

Of particular concern are specialist programs. For example, reggae lovers are a tiny minority in most stations' audiences. Some listeners are ambivalent about reggae, and some will actively dislike it, but everyone is probably satisfied if you play one to two reggae songs per hour. The lovers, the ambivalent, and the haters all accept their lot.

However, when you decide on a reggae hour, you actively tell a significant portion of your audience to tune out. Research carefully before committing to a specialist show.

You may be actively chasing listeners away. "Mixes" are the ultimate specialist show, and the only true lovers of those mixes are almost certainly the DJ and night owls who live their lives in clubs.

Despite my concerns about "specialist programming", Brad March believes they have a place: *A break in the regular music format can help prevent listeners from experiencing format fatigue. The SCA RnB Friday is a good example of this. Or the "Classic 9 at 9" feature at 9 am is a strong feature or benchmark supporting the station's music position.*

While I'm generally not a fan of "block" programming, it can be beneficial as long as it supports the station's music or target demo.

Additionally, there are times when it's necessary to create a program or show for sales/sponsor's purposes. This is fine, again as long as it aligns with the station's target demographic, music format, and fits the brand. Additionally, Australian radio has done a great job creating shows and programs for talent development.

For instance, Triple M currently (July 2025) has podcasters Lu and Jarch on air, and they are possibly trialing and testing them on a peak time show in the future. This is a good use of "block" programming and airtime.

I can't see block programming like a "Jazz" show on FM radio in Australia. It is too niche. It may work on digital, though.

If you insist on specialist shows, the perfect tool is the **clock constraint rule**. For example, you might want to play only Gospel songs during a Sunday morning Gospel program, or only Jazz songs during a specialist Jazz program on Thursday evenings.

Songs can be assigned a Genre (Jazz) and a Sound Code (Big Band, Dixie, Smooth, Sax, Piano).

Thus, you could assign Jazz (Genre) constraint to every position except those that specifically must be Big Band Jazz (Sound code).

You could also insist that no reggae songs play during specific periods of the day.

Or you might use a clock constraint rule based on energy. This rule enables you to control the energy of any specific position or hour.

For example, the required general energy for the station might be different from the energy needed for a late Friday night "party" time versus 3.00 am, when you might prefer a more laid-back atmosphere.

Ideally, we are looking for light and shade, hills and valleys. If you do not want continuous high-level energy, give us a break occasionally with something slightly less energetic. The same goes for Tempo.

How do GSelector/GLinker Schedule

First, we need to address the elephant in the room. Soon, we will see AI based music and programming. But, the philosophies and practices we believe in will hold true. It remains to be seen just how effective AI scheduling will be. I eagerly await the arrival of RCS's first AI GSelector releases.

What follows is primarily from the GSelector HELP; no other reference is more suitable.

The primary rule that every song in a category should have equal plays to its peers is addressed by GSelector, which reviews each song in the most-rested stack and attempts to schedule the song at the top of, or as close to the top of, the stack first.

The "STACK" order is a sequential list of a category: the item at the top of the list is the first contender for scheduling, the second item in the list is the second contender, the third item in the list is the third contender, and so on.

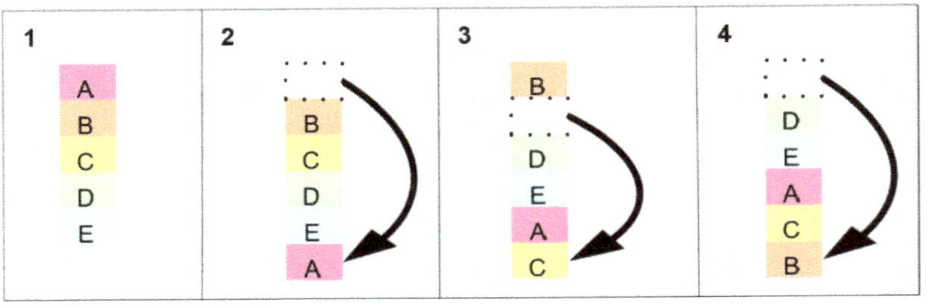

In Stack #1: we have a category stack order A to E.

In Stack #2: When the stack is considered for scheduling, if A is satisfactory, meets all the required rules, and breaks none (or the fewest breakable rules), then it is scheduled in that position and moved to the bottom of the stack.

In Stack #3, B is first considered for scheduling but fails an unbreakable rule. However, C fails no rules and is therefore scheduled and moved to the bottom of the stack.

In Stack #4, B is again considered for scheduling, meets all the requirements, and is therefore scheduled and moved to the bottom of the stack.

You will note that the stack is now "out of sequence". This explains why some songs might become candidates for scheduling sooner than expected, even before their projected natural turnover.

GSelector has a way of dealing with songs that are frequently skipped, a phenomenon known as **Airplay Starvation**.

This forces a song that has passed a defined threshold of not being played to be played. Remember, every song in a category deserves to have equal plays to its peers.

As each position is examined and scheduled, GSelector undergoes a multi-phased process. GSelector and GLinker employ top-down priority scheduling.

This allows the system schedulers to quickly fill every position with the "best" song or link, following the Stack order, the Goals and rules you use, and their priorities.

When examining items for possible use in a given schedule position, the scheduler works with the policy's rules, settings, and priorities for the time block for that position.

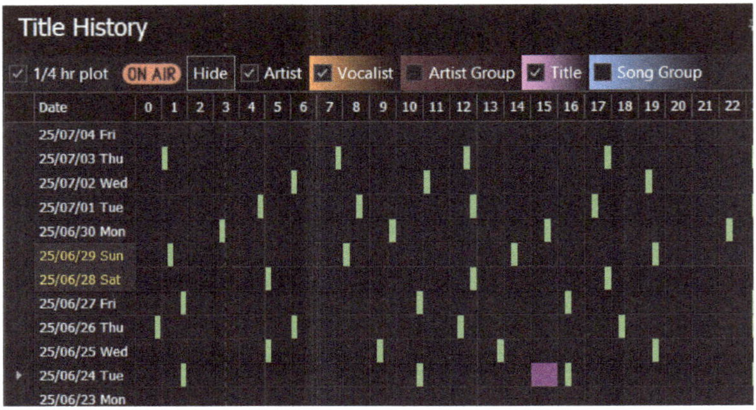

A typical map of the history of when a song has been scheduled. Note the nearly perfect natural rotation, demonstrated by the "ladders" showing each play seen. The expectation for this station is a TSL of one hour per day.

The scheduler starts with the first item in the stack of the appropriate category and level. It tests that item for your rules, starting with unbreakable rules and working down the priority list. The scheduler tests the item against every prioritized rule. If the item does not violate any rule, it is scheduled.

If it violates an unbreakable rule, the scheduler steps past that schedule position and proceeds to the next item in the stack. If the item fails a breakable rule, the system adds the song ID or link ID and the priority number of the failed rule to a temporary list, then proceeds to the following item in the stack.

Examining items and testing rules continue until an item is scheduled or until all items within the search depth have been reviewed.

If no items are on the temporary list when the search depth has been reached, the position is left unscheduled. If there is only one item on the temporary list, it is scheduled.

If there are more than one item on the temporary list, the one that violates the rule with the lowest priority, i.e., the item with the highest priority number, is scheduled. If the temporary list contains multiple items that violate the rule with the lowest priority, the most-rested items are scheduled.

The scheduler then moves on to the next position in the schedule.

This scheduling process means that the audit trail will never reference a rule being dropped. If an item fails a higher-priority rule, there's no need to test it again for a lower-priority rule.

Since rule testing is now a top-down process, a given rule is no longer dropped as a specific step of the scheduling process. However, the notion that a rule is ignored or dropped remains alive.

The system will ignore or drop one or more lower-priority rules to protect one or more rules at a higher priority.

GSelector will examine the first song at the top of the stack and check if it respects all the unbreakable rules set for that category.

If the song breaks any of the unbreakable rules set for that category, it will not be scheduled, and GSelector will move on to the next song in the stack for review.

Once GSelector finds the highest song in the stack that does not break any Unbreakable rules, it will submit that song to the Breakable rules testing, starting with the least important rule at the bottom of the Unbreakable rule list.

If a song does not break these rules, then GSelector will schedule it. If it does break the lowest rule, GSelector will move on to the next song in the stack.

Eventually, a song will be found that does not break Unbreakable rules and breaks the fewest Breakable rules; and it will be scheduled.

It is possible that GSelector cannot find a satisfactory song above the Search Depth that satisfies the rules and Goals, and in this case, the position will remain unscheduled.

Unscheduled positions in GSelector happen very seldom. If they do happen, you almost certainly have a problem with the database balance clashing with your expectations defined by the category composition, goals, and legacy rules you have established.

This is a general overview of the scheduling process. It explains how GSelector's approach to scheduling is unique and designed to provide the best possible scheduling solution, rather than the least bad solution.

The measurement for this is found in the Log's **Score** column in the generated Log.

100 is when GSelector found it easy to schedule; anything below 80 suggests you need to review the goals, rules, balance of all the attributes or special clock constraints you have imposed.

Whether the demand from each clock position is determined by 'regular' clock positions, or by using rolling clock grids or clock constraints, GSelector seeks the optimum choice for each position. Song scheduling occurs in Daypart order, as set in the Dayparts.

So, you can schedule the morning drive first, followed by the afternoon drive, and then the morning, afternoon, evening, and overnight drives—your choice.

"Rolling Clocks" is a feature that reduces the predictability of your scheduling, allowing you to set up a variety of Feeder clock sequences. For example, you can create sequences such as category A, B, B, D, A, D, and category B, D, A, A, C, B.

Now, you set up a master clock that calls for these clock sequences in a rolling sequence.

Equally, you could have a Linker rolling clock set, which is called for by the master clocks.

In Phase 1 of GSelector scheduling, faster-turnover and slotted song category groups are scheduled in the pass-order sequence in the category group options window. These song category groups generally turn over so fast that daypart rotation is not a scheduling issue.

Phase 2 consists of three steps: pre-planning, scheduling, and final optimization.

In the pre-planning step, GSelector temporarily schedules the remaining song category groups and link categories according to the daypart order set in the Dayparts Definition window.

During the pre-planning stage, all goals and rules are ignored. Songs are allocated to dayparts in the next most rested order.

This distributes the most due Songs (in most-rested order) across the schedule, approximately where they should be by daypart.

Step 2 is the scheduling process, which considers all the goals and other settings. When the first part is scheduled, the pre-planned songs for that part are unscheduled and available for scheduling.

Note that all songs in subsequent and further dayparts remain pre-planned until GSelector starts scheduling the next daypart, when that daypart's pre-planned allocation of songs and links is released for scheduling, and so on.

Step 3 is to revisit the finished schedule to optimize it. Working in turn in each daypart across the scheduled period, songs are shuffled within their respective song category groups in each scheduled position to find the combination of songs with the best possible scores for each position, constantly respecting goals and settings.

Different conditions apply to Links. Specific Fixed Links in Clocks are placed in the schedule first when GSelector reads the Clock before the schedule process begins.

Then, link category positions are scheduled before songs, in the order set in the Link Goals Categories window.

NOTE: You can split the scheduling process into separate song and link scheduling runs. Refer to the 'Scheduler Window' topic for help on scheduling songs or links separately. This will allow you to schedule links after songs, which allows song-link texture segue bans to work more efficiently.

Finally, **Related links** are scheduled after the songs. Read about Related Links in Help. This is an excellent capability; you could associate a link advertising a live tour with the artist. Every time the Artist is scheduled, the related link is also automatically scheduled.

By default, GSelector allows the total number of songs in any category group to be searched using regular scheduling. This means the scheduler will consider all songs in the categories in a Group. Under most circumstances, a reasonable, well-balanced library will spread items according to the natural demand of the library.

However, there is a risk, especially with smaller, high-rotation categories, that this may result in less-than-optimal rotation, favoring easy-to-schedule songs while underutilizing or restricting harder-to-schedule songs.

So, depending on your library, there may be situations in which you wish to control rotation more precisely.

By default, all search depth (depth %) values are set to 100%, allowing GSelector to consider all songs in the categories for that category group. You can choose to reduce this value., it certainly effects how often a song is scheduled.

While search depth values are expressed as percentages, GSelector dynamically adjusts the search depth setting to the percentage equivalent of a single song in small categories within a category group.

The value automatically adjusts internally, regardless of the number of songs in the group, unless the group contains more than 100 songs.

This effectively allows you to use a slotted approach (start at the top of the stack, work through to the bottom, then start at the top again). In addition, to avoid the same sequence appearing in the same hours, you can **Kick.** Read the manual.

GSelector schedules each category one at a time. It starts with the category at the top of the scheduler's pass order.

Unless you have a prioritized daypart for scheduling, GSelector then examines the clock in the first hour and searches for all songs or links in this pass order, scheduling each song for the entire day and through to the end of the expected period.

Then, GSelector starts again for the scheduled period with the following category in the pass order.

Search depth is one of the most critical scheduler policy settings because of its direct role in scheduling.

When filling a scheduled position, the system always attempts to find the best item within your specified search depth.

Search depth is dynamically altered, depending on requirement. But, it never searches more deeply into a category or level than you allow.

TIP:

In the categories tab of 'policy parameters' for GSelector and 'policy parameters' for GLinker, you can adjust several search depth options. These affect how the associated system scheduler determines search depth for various items.

The search depth setting significantly influences how soon an item in a category or category level is likely to repeat. Say that you allow GSelector to search two songs deep into a level containing nine songs.

Furthermore, our clocks are designed so that there is one clock request per hour for each level. If precisely rotated, each song in this level would repeat every nine hours.

In this example, our search depth indicates that we will allow a song in a level to repeat every seven hours, if necessary, to prevent a rule from being violated.

The search depth setting implies that we feel it is better to have a song in the category/level repeat in seven hours than to violate a rule.

Some items are easier to schedule than others. If the search depth is too deep, the items that are "easy" to schedule will get a larger share of airplay than those that are harder to schedule.

If the search depth is too shallow, the scheduler may be unable to find an item that meets your goal or rule requirements.

The GSelector and GLinker schedulers each work on one category at a time, and the order in which categories are scheduled is a critical consideration set in the category **Pass order.** Also, it is possible to set an order for which dayparts get scheduled in which order.

For example, schedule the morning and evening drive shows first, then proceed with the rest of the day.

This ensures that your best songs can play in your most important dayparts first.

There are fewer potential rule violations when the first two or three categories are scheduled simply because there are fewer existing items in the schedule that cause conflicts.

The system will not have to dig as deep when scheduling categories in the early phases of the scheduling pass order.

This is especially true of small number categories. Schedule them first. As more categories are scheduled, more rule conflicts are likely to arise.

The items in the planned categories can now present problems for those being considered. A good rule of thumb is that the later a category is scheduled, the higher its search depth will need to be.

GSelector's search depth is dynamic. It deepens as required. You're trying to achieve a happy compromise with a search depth setting.

If the setting is too small, the system may struggle to find items that satisfy your rules.

If your search depth is too large, you will get uneven rotations because the system will favor those items that are "easy" to schedule. The items that are "hard" to schedule will get skipped.

TIP:

If you wish to precisely rotate a small category/level with a relatively quick turnover, assign a priority list to this category, which has no rules other than clock goals.

Set the category to the first position of the pass order and set its search depth to ONE.

This approach will provide a rotation in which every item in the category or level will be laid out in the schedule in precise stack order. Read about **Kicks**.

This is just a fraction of what GSelector can achieve. As you do, you will find that the more you use GSelector, the more you will rely on the Help function to locate a capability that you are convinced must be there somewhere.

So, despite real men never reading manuals, be assured that reading the GSelector online context-sensitive Help will always empower you.

As difficult as it is to close, it must occur at some point. Here is something from a well-known but non-radio personality that I have chosen to modify slightly. Enjoy, have fun, be passionate.

And always remember that working in radio isn't a job – it's simply the most fun you can legally have in a career.

Media Monitors/Aircheck Data for Research

A fascinating, little-known fact is that the RCS's Aircheck/Media Monitors off-air monitoring software evolved from Cold War-era software technology, which enabled the identification of Soviet submarines' propeller sound prints—a significant tactical advantage to the Western world.

https://www.mediamonitors.com/

I warn you against following your competitors; they are not your target audience. However, handy information is available to track exactly what they are doing compared to you. Knowing what your competitors are doing is a significant strategic advantage. Aircheck/Media Monitors, RCS's off-air monitoring services, are one of the most extensive and capable products available to a radio station worldwide. Without it, or a similar product in your arsenal, your potential for success is compromised.

Real World Research

This data is from one of the major stations in Namibia, and yes, we have permission to show it.

In Windhoek, Namibia, this station is unashamedly oriented towards Afrikaans-speaking people. However, one of the surprising outcomes of this research has led the station to retain its "Afrikaans" character while becoming a bilingual station (offering both Afrikaans and English presenters), demonstrating its commitment to serving its diverse audience.

This data was achieved using RCS's "Test-all-Media" and a heavily promoted "Join our programming consultants' team" promotion, which directed participants to a microsite on the station's website. Each month, a substantial prize was offered for participation. The fundamental aspect of bringing the listener into the family was proven.

Primarily, the station was dedicated to understanding its listeners, particularly the music that appealed to its P1 listeners. Perceptual research was also conducted to identify their existing and potential listeners.

The music research had the advantage of dealing with known participants over several years. Over time, the station began to gain very granular and deep insights into what kind of music appealed to their P1 listeners and precisely what the life cycle of each song was for each specific listener. Invaluable data.

This was an ongoing campaign with monthly music life-cycle research, and during COVID-19 in 2020, quarterly or even sometimes monthly. The true value of this data is that one digs deep into the listener's favorite music and their tastes, and whether our rotations and playlisted songs are acceptable or not.

Perceptual research campaigns.

The perceptual research was also focused on arming their sales teams with compelling numbers to justify their major clients' spending on their station and attract other potential clients.

The results of each project were extensively presented and explained to management, programming, sales, and all other interested parties.

The station conducted monthly research to equip its salespeople with compelling data to support their pitches. The numbers often closed the deal.

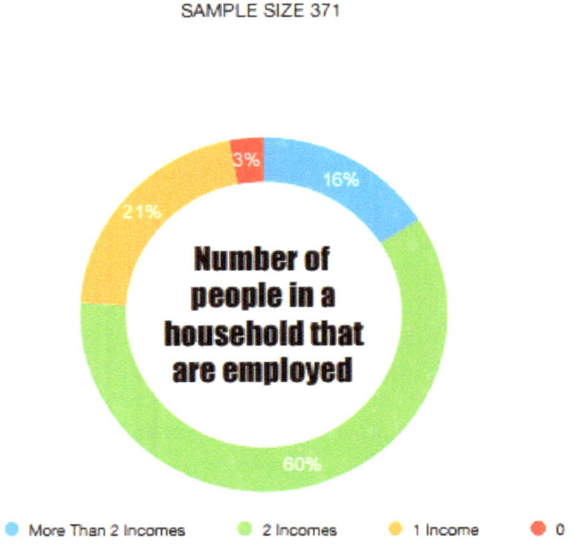

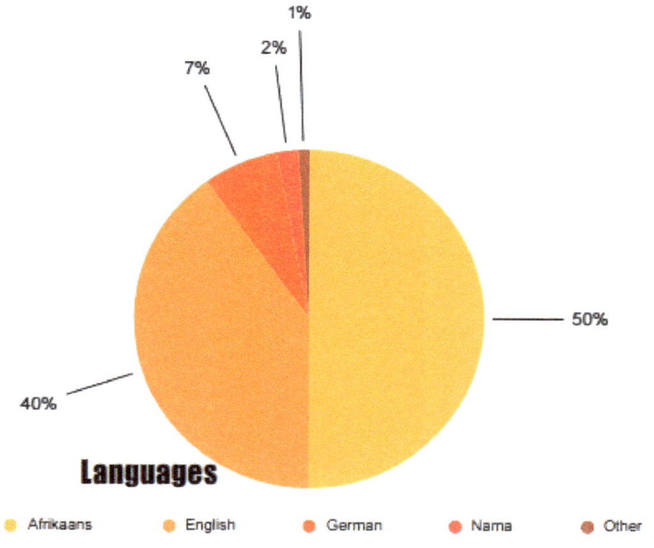
1%
2%
7%
50%
40%
Languages
Afrikaans
English
German
Nama
Other

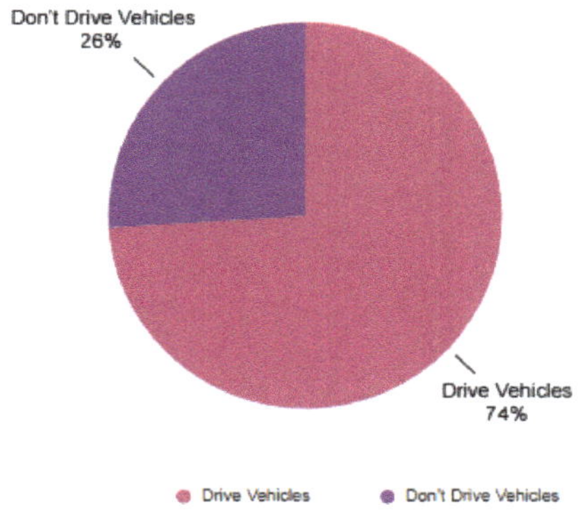
Don't Drive Vehicles
26%
Drive Vehicles
74%
Drive Vehicles
Don't Drive Vehicles

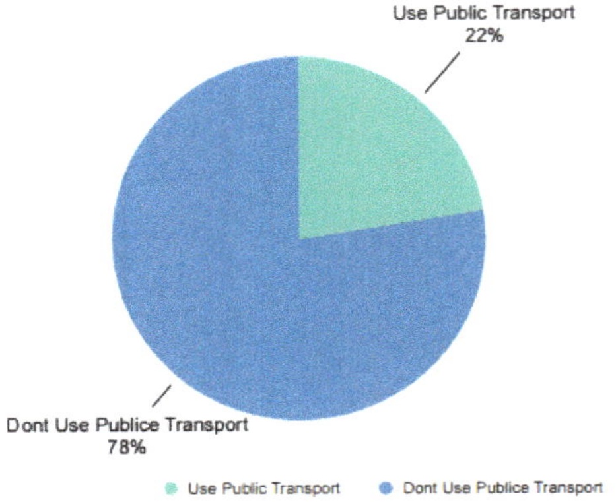

Use Public Transport
22%
Dont Use Publice Transport
78%
Use Public Transport
Dont Use Publice Transport

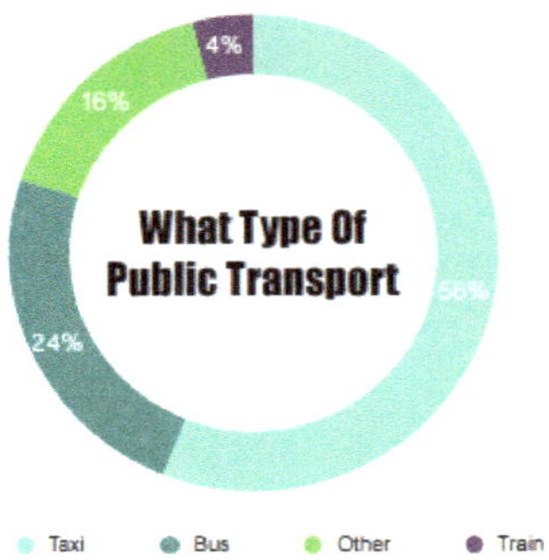

What Type Of Public Transport
4%
16%
24%
58%
Taxi
Bus
Other
Train

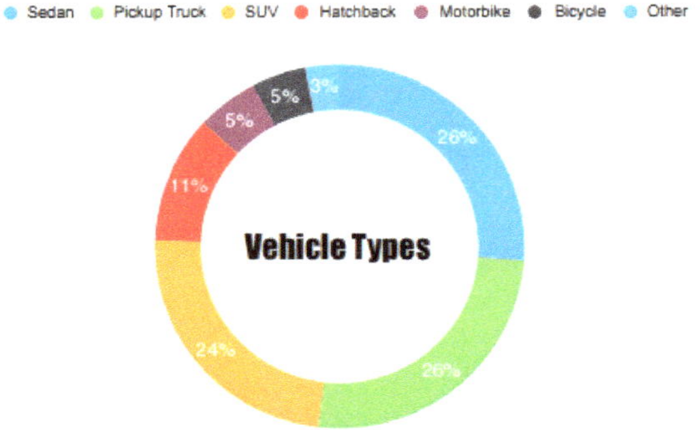

Sedan
Pickup Truck
SUV
Hatchback
Motorbike
Bicycle
Other
Vehicle Types
5%
5%
3%
26%
11%
24%
26%

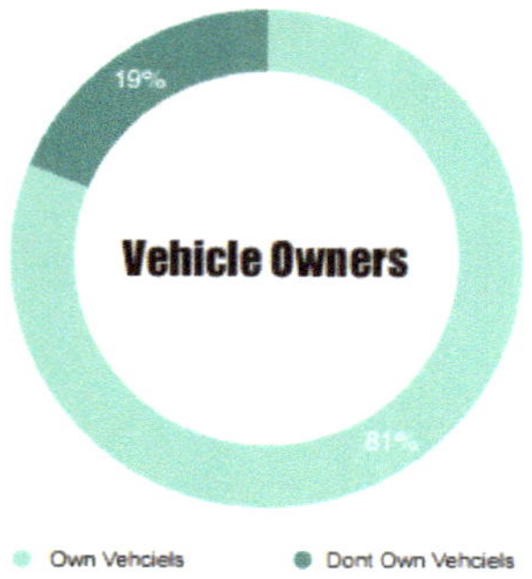
19%
Vehicle Owners
81%
Own Vehciels
Dont Own Vehciels

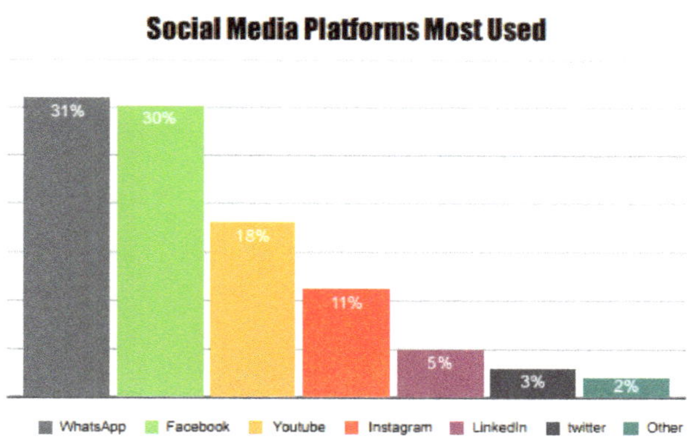
Social Media Platforms Most Used
31%
30%
18%
11%
5%
3%
2%
WhatsApp
Facebook
Youtube
Instagram
LinkedIn
twitter
Other

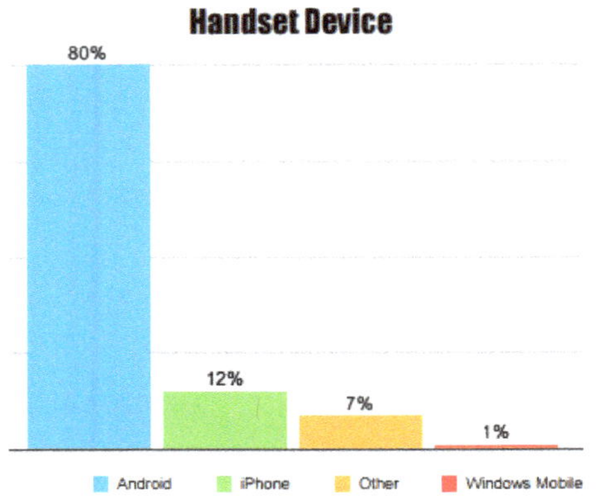
Handset Device
80%
12%
7%
1%
Android
iPhone
Other
Windows Mobile

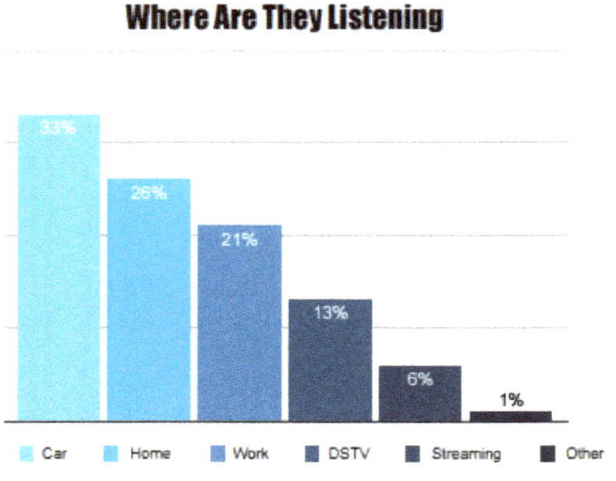
Where Are They Listening
33%
26%
21%
13%
6%
1%
Car
Home
Work
DSTV
Streaming
Other

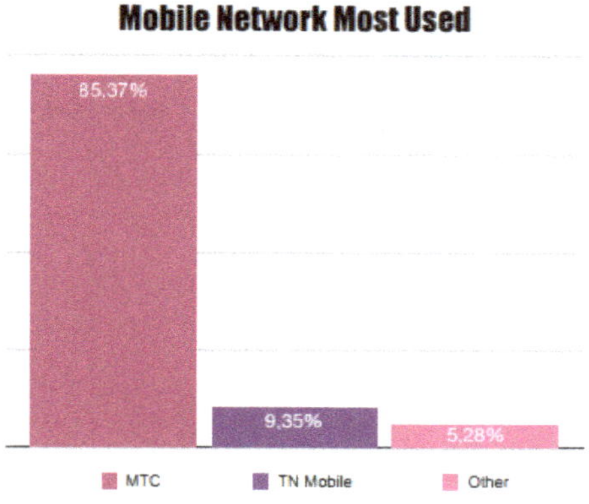

And this is also an opportunity to investigate our general "sound" of the station.

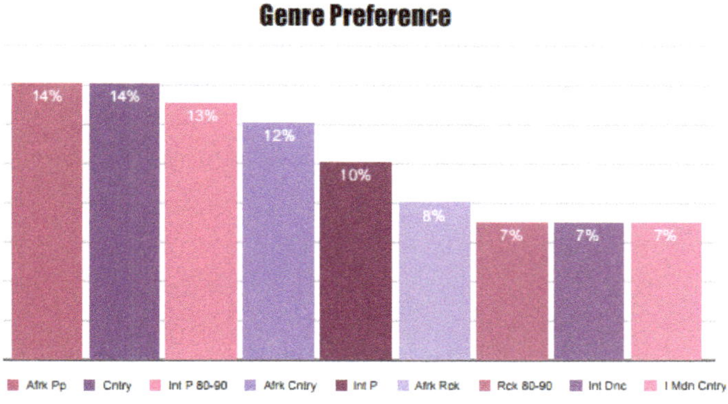

It is also important to know where your listeners are based.

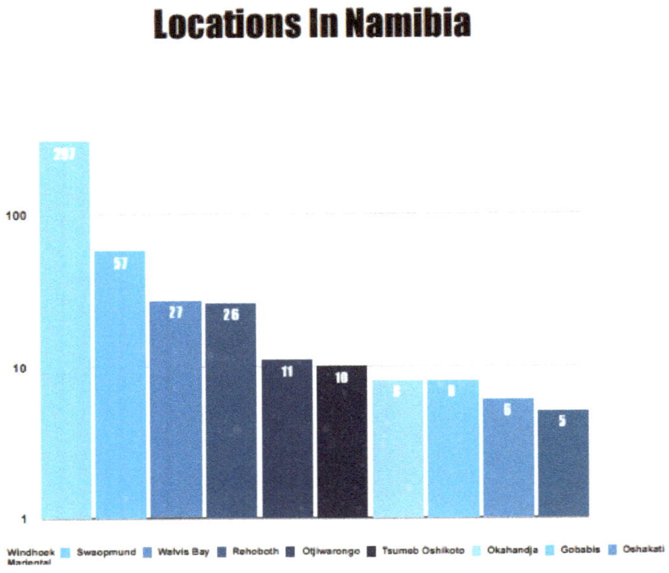

Africa has a huge diaspora. Where are they listening to us?

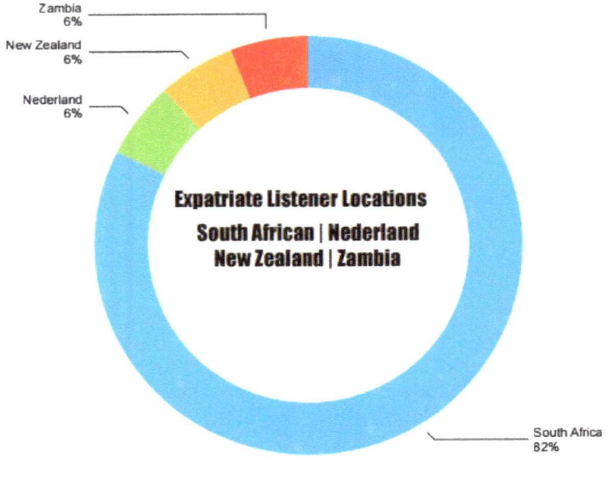

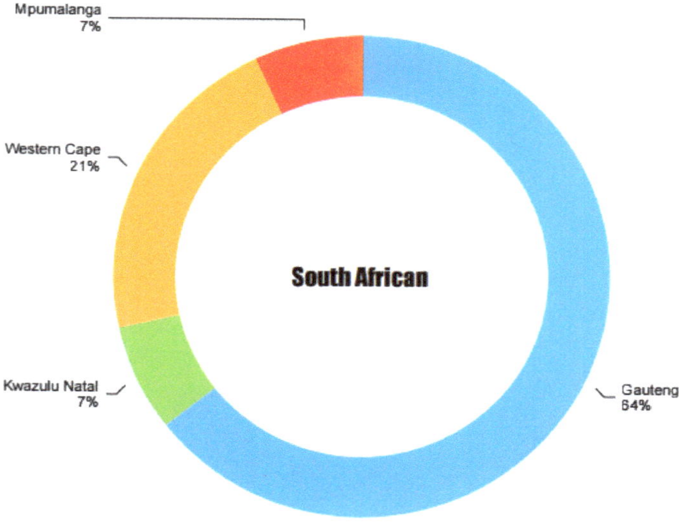

Trying to get a sale from a pet store? Here is valuable data to close the deal.

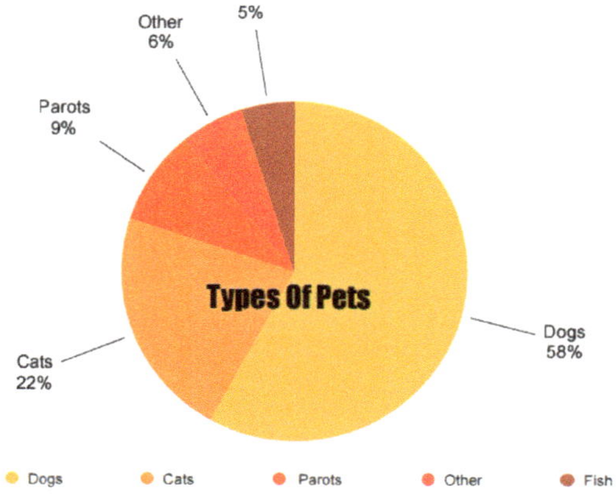

We know that newspapers are widely read. But which ones?

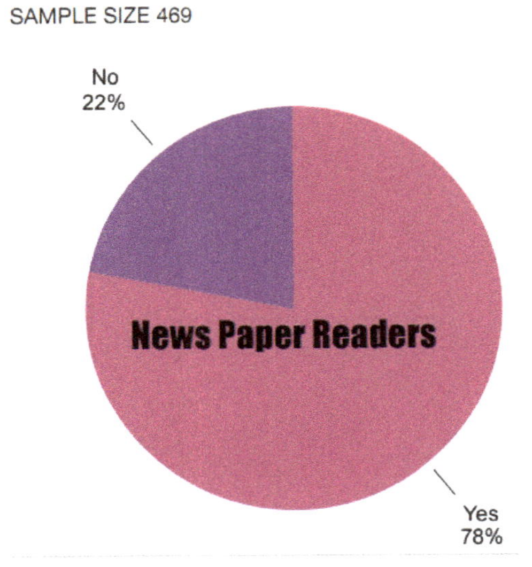

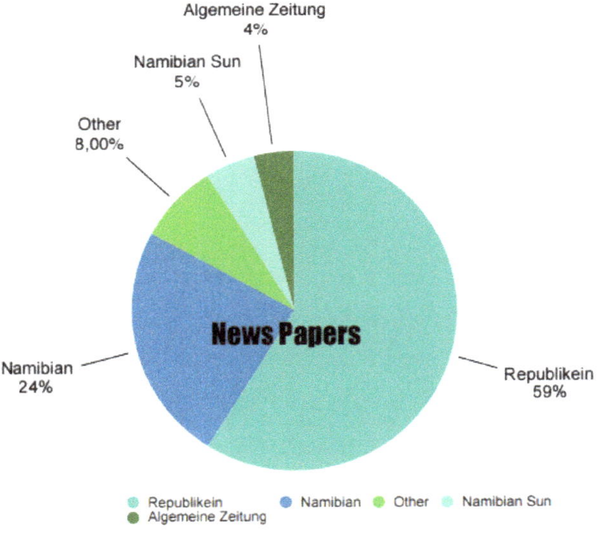

Where are our listeners mainly seeking News

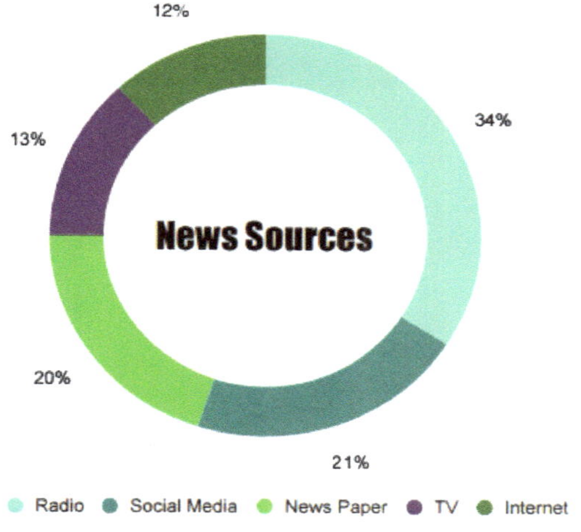

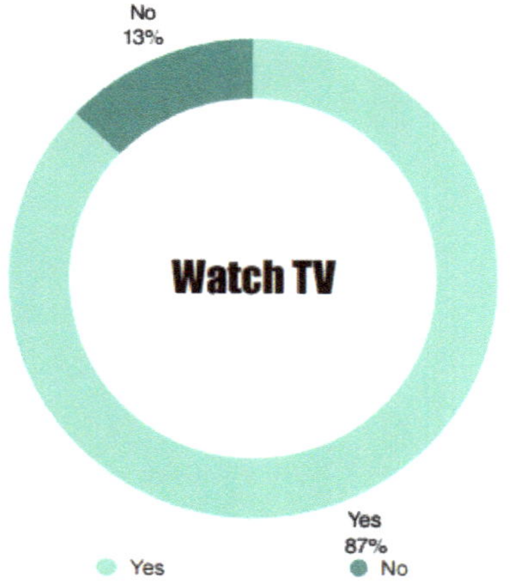

In terms of potential TV channel clients, who should we target?

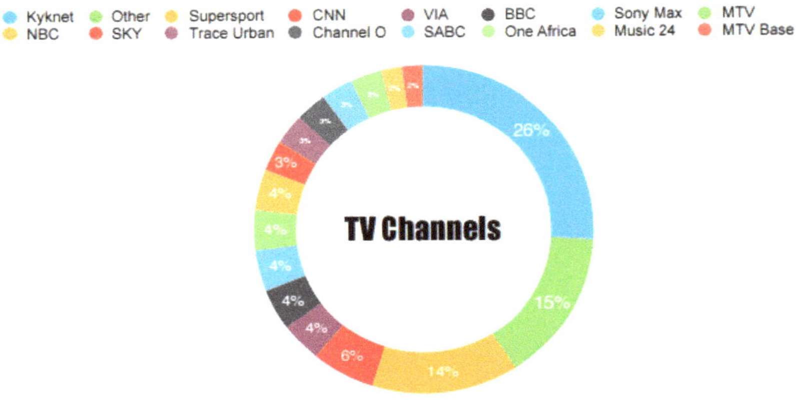

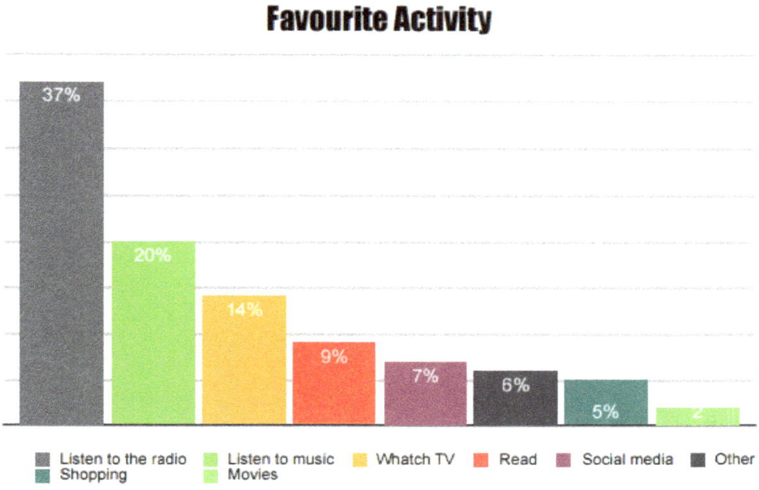

When are our highest listeners, by daypart?

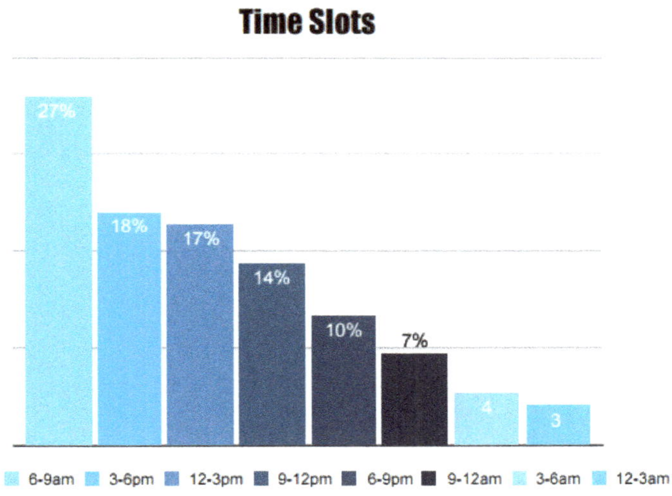

Are “Special Deals” offers attracting customers to the marketer?

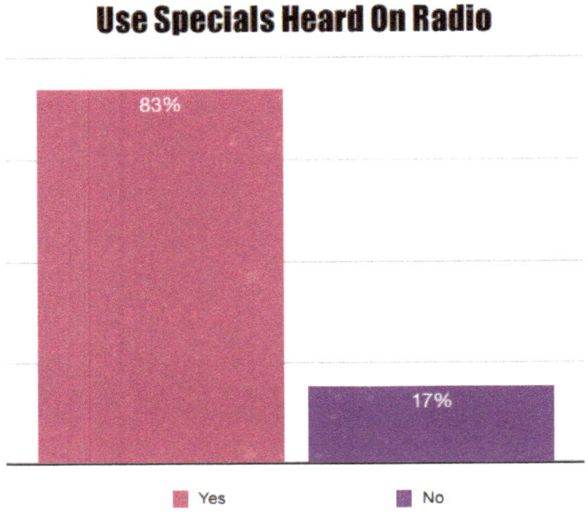

This data was required when we were approaching several insurance companies. Those at the top of the pile were delighted by our listeners’ responses and signed up for further campaigns. Those at the bottom were less delighted. However, these numbers provided a powerful tool to close sales for several insurers.

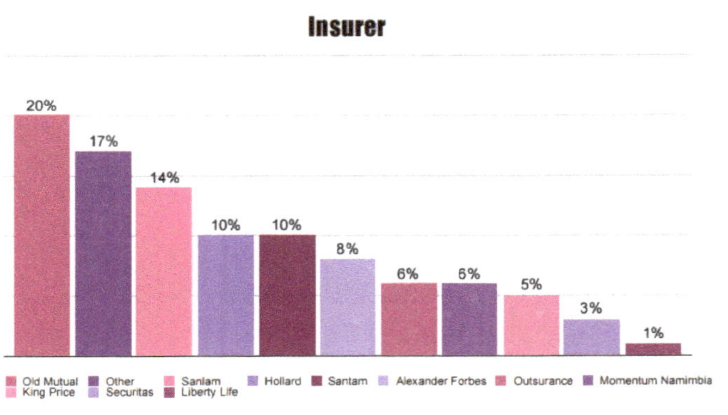

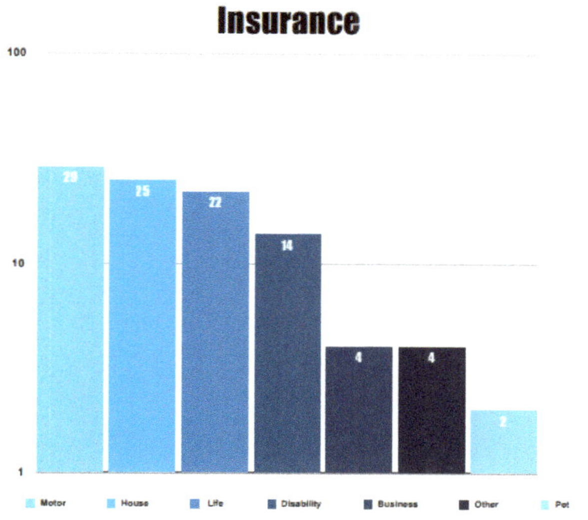

The fact that a quarter of our listeners had no Medical insurance was a powerful motivator to one Medical Aid company we were trying to sell to. These numbers helped close the deal.

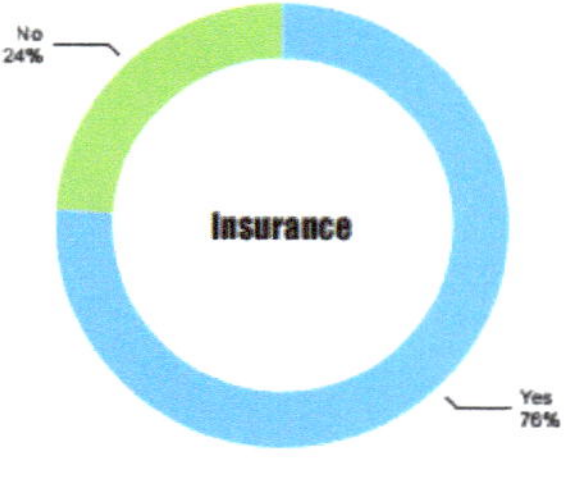

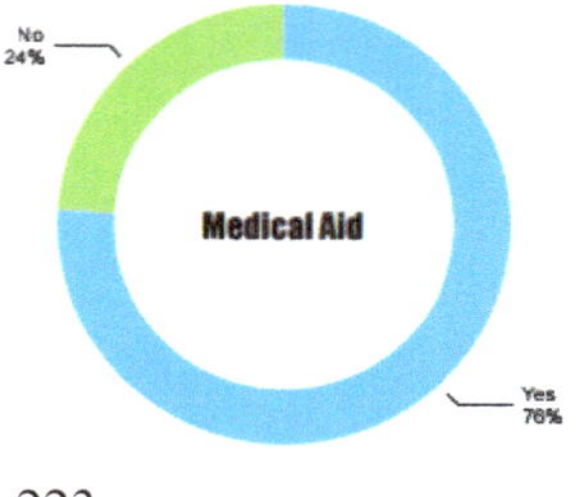

Tracking the listeners' acceptance of our playlist and format is vital. Comparing week to week, we can now see exactly where a song's popularity stands on the bell-curve, and we can adjust our playlist appropriately.

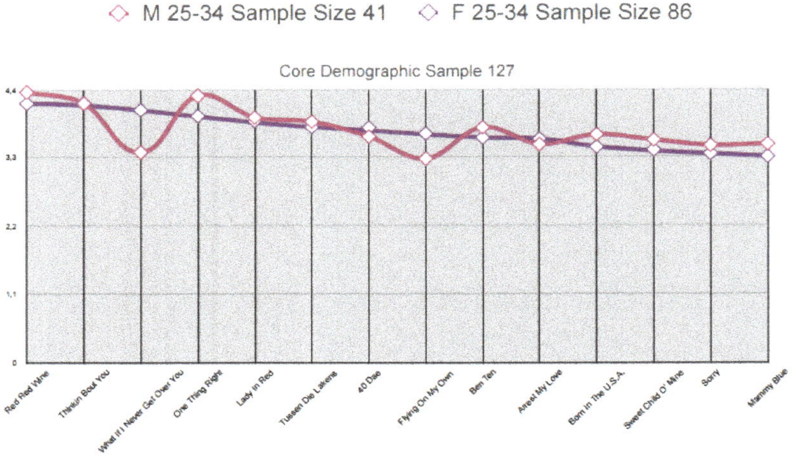

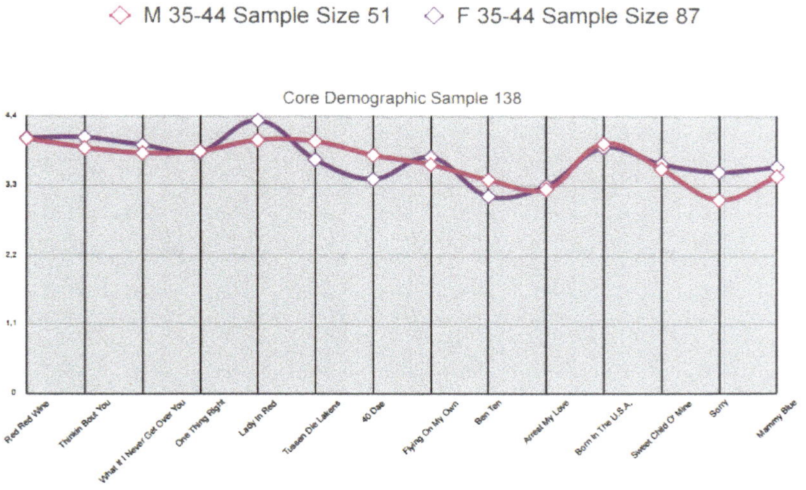
M 35-44 Sample Size 51
F 35-44 Sample Size 87
Core Demographic Sample 138
4,4
3,3
2,2
1,1
0
Red Red Wine
Thinkin Bout You
What If I Never Get Over You
One Thing Right
Lady In Red
Tussen Die Lakens
40 Dae
Flying On My Own
Ben Ten
Arreal My Love
Born In The U.S.A.
Sweet Child O' Mine
Sorry
Mammy Blue

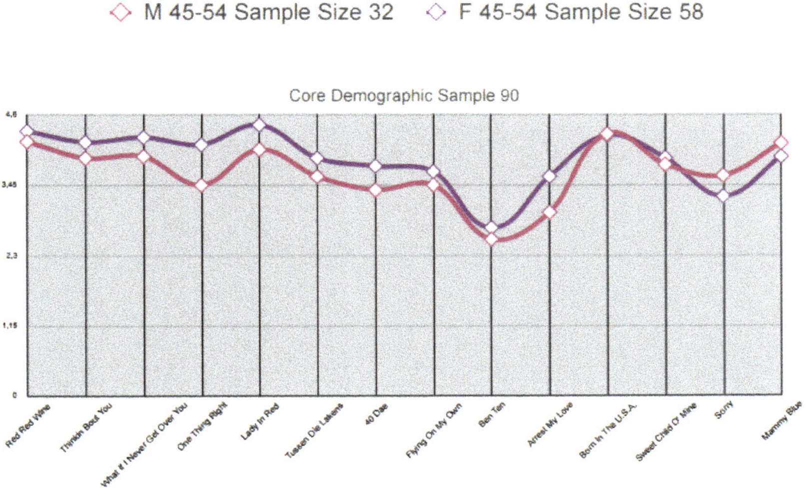
M 45-54 Sample Size 32
F 45-54 Sample Size 58
Core Demographic Sample 90
4,6
3,45
2,3
1,15
0
Red Red Wine
Thinkin Bout You
What If I Never Get Over You
One Thing Right
Lady In Red
Tussen Die Lakens
40 Dae
Flying On My Own
Ben Ten
Arreal My Love
Born In The U.S.A.
Sweet Child O' Mine
Sorry
Mammy Blue

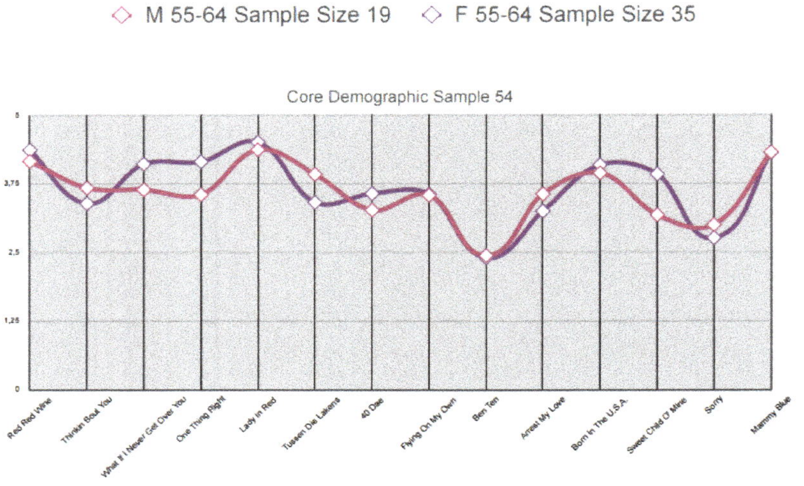
M 55-64 Sample Size 19
F 55-64 Sample Size 35
Core Demographic Sample 54
5
3,75
2,5
1,25
0
Red Red Wine
Thinkin Bout You
What If I Never Get Over You
One Thing Right
Lady In Red
Tussen Die Lakens
40 Dae
Flying On My Own
Ben Ten
Arrest My Love
Born In The U.S.A.
Sweet Child O' Mine
Sorry
Mammy Blue

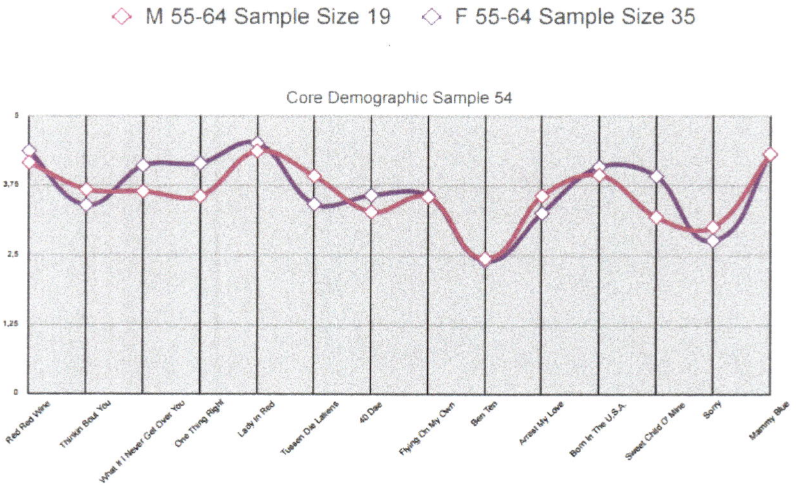
M 55-64 Sample Size 19
F 55-64 Sample Size 35
Core Demographic Sample 54
5
3,75
2,5
1,25
0
Red Red Wine
Thinkin Bout You
What If I Never Get Over You
One Thing Right
Lady In Red
Tussen Die Lakens
40 Dae
Flying On My Own
Ben Ten
Arrest My Love
Born In The U.S.A.
Sweet Child O' Mine
Sorry
Mammy Blue

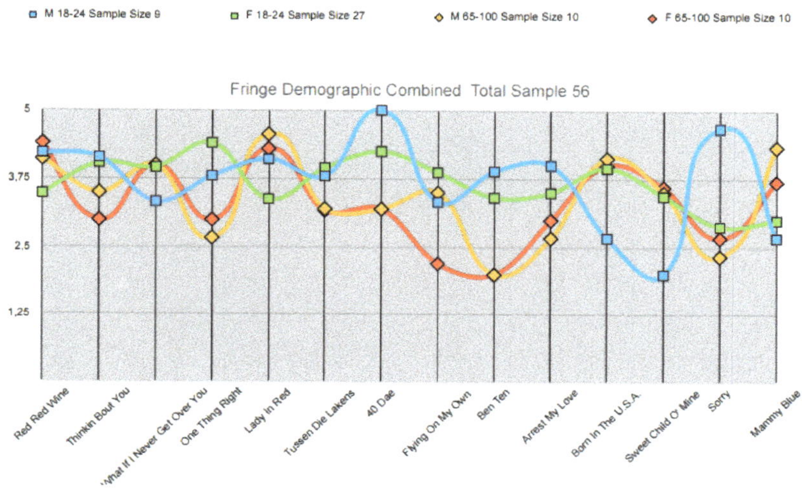

This is the raw data from which the various graphs are built.

	18-24 (36)		25-34 (127)		35-44 (138)		45-54 (90)		55-64 (54)		65-100 (20)	
	Gender: Female (27)	Gender: Male (9)	Gender: Female (86)	Gender: Male (41)	Gender: Female (87)	Gender: Male (51)	Gender: Male (32)	Gender: Female (58)	Gender: Male (19)	Gender: Female (35)	Gender :Female (10)	Gender: Male (10)
Title	Mean Fam	Mean Fam	Mean Fam	Mean Fam	Mean Fam	Mean Fam	Mean Fam	Mean Fam	Mean Fam	Mean Fam	Mean Fam	Mean Fam
Mammy Blue	3.00	2.67	3.30	3.50	3.59	3.44	4.13	3.91	4.32	4.31	3.70	4.33
Sorry	2.88	4.67	3.35	3.48	3.51	3.07	3.60	3.26	3.00	2.76	2.67	2.33
Sweet Child O' Mine	3.44	2.00	3.41	3.57	3.64	3.56	3.78	3.89	3.18	3.92	3.60	3.50
Born In The U.S.A.	3.95	2.67	3.46	3.66	3.90	3.96	4.28	4.26	3.94	4.09	4.00	4.13
Arrest My Love	3.50	4.00	3.59	3.50	3.29	3.24	3.00	3.59	3.56	3.25	3.00	2.67
Ben Ten	3.41	3.89	3.61	3.78	3.13	3.39	2.56	2.75	2.44	2.40	2.00	2.00
Flying On My Own	3.87	3.33	3.68	3.27	3.75	3.63	3.45	3.67	3.54	3.56	2.20	3.50
40 Dae	4.25	5.00	3.78	3.63	3.40	3.78	3.37	3.76	3.27	3.57	3.20	3.20
Tussen Die Lakens	3.95	3.80	3.79	3.88	3.71	4.00	3.59	3.89	3.92	3.41	3.17	3.20
Lady In Red	3.38	4.11	3.89	3.96	4.33	4.02	4.03	4.43	4.37	4.51	4.30	4.56
One Thing Right	4.40	3.80	3.98	4.31	3.83	3.84	3.45	4.11	3.55	4.15	3.00	2.67
What If I Never Get Over You	3.96	3.33	4.08	3.83	3.94	3.81	3.92	4.23	3.64	4.11	4.00	4.00
Red Red Wine	3.48	4.22	4.19	4.37	4.05	4.04	4.16	4.33	4.16	4.37	4.40	4.11
Thinkin Bout You	4.04	4.14	4.19	4.19	4.06	3.89	3.89	4.15	3.67	3.39	3.00	3.50

Note: We apologize for the subpar quality of the graphics, they do not convert well to print.

www.ingramcontent.com/pod-product-compliance
Ingram Content Group UK Ltd.
Pitfield, Milton Keynes, MK11 3LW, UK
UKHW022021041026
15591UKWH00012B/429

9 781037 001345

Maths, Art, or Magic?

Radio station scheduling and programming

Third edition 03/01/2026
ISBN 978-1-0370-0134-5

First Published in 2025
PR. Prior